Carbon Nanostructures

For further volumes:
http://www.springer.com/series/8633

César Avellaneda
Editor

NanoCarbon 2011

Selected Works from the
Brazilian Carbon Meeting

 Springer

Editor
César Avellaneda
Centro de Desenvolvimento Tecnológico
Universidade Federal de Pelotas
Pelotas
Brazil

ISSN 2191-3005   ISSN 2191-3013   (electronic)
ISBN 978-3-642-42706-0   ISBN 978-3-642-31960-0 (eBook)
DOI 10.1007/978-3-642-31960-0
Springer Heidelberg New York Dordrecht London

Preface

This book contains a brief overview of selected works presented at the "X Meeting on Diamond, Amorphous Carbon, Nanotubes and related Materials" held at Pelotas-RS- Brazil, June 2011. This book remarks the advances occurring since the past decade to date and presents good perspectives of future research in the carbon nanotube field.

On behalf of the organizer committee, we would like to thank the authors who produced excellent chapters from which the readers interested in carbon nanotubes and related materials will greatly benefit, and to Springer-Verlag for cooperating with us in implanting this project. We also acknowledge the financial support from the Brazilian funding agencies CNPq and FAPERGS.

Pelotas, July 16, 2012   Prof. Dr. César O. Avellaneda

Contents

Review of Field Emission from Carbon Nanotubes: Highlighting Measuring Energy Spread

M. H. M. O. Hamanaka, V. P. Mammana and P. J. Tatsch

Abstract This paper is a review of the research on field emission properties of carbon nanotubes (CNTs), the basic properties of CNTs, the main emission properties with highlighting in energy spread and the work done in applying CNTs for field emission microscopy (FEM). In this work there are explanations about the density of states (DOS) of the conduction electrons responsible for the emission; comparison of the characteristics of CNTs emission from single nanotube or films; comparison of the different types of electron sources and the introduction of CNTs electron sources applying in retarding field analyzer (RFA).

1 Introduction

In the Division of Information Displays of CTI several topics on displays are investigated, for instance: Liquid Crystal Displays (LCDs), Organic Light Emission Displays (OLEDs), Polymer Stabilized Cholesteric Texture Displays (PSCT), Polymer Dispersed Liquid Crystals Displays (PDLCs) and Field Emission Displays (FEDs). Although FEDs are not playing an important role in the display world anymore, the range of applications of carbon nanotubes CNTs) is still growing, as will be described hereafter. We are mainly interested in the field emission properties of CNTs in applications such as electron microscopes and other electron optic devices.

M. H. M. O. Hamanaka (✉) · V. P. Mammana
Centro de Tecnologia da Informação Renato Archer – CTI, Campinas-SP, Brazil

M. H. M. O. Hamanaka · P. J. Tatsch
Universidade Estadual de Campinas – UNICAMP, Campinas-SP, Brazil

C. Avellaneda (ed.), *NanoCarbon 2011*, Carbon Nanostructures,
DOI: 10.1007/978-3-642-31960-0_1, © Springer-Verlag Berlin Heidelberg 2013

Fig. 1 Longitudinal cross-
section of a CRT fluorescent
display with a FE cathode
made of carbon [71]

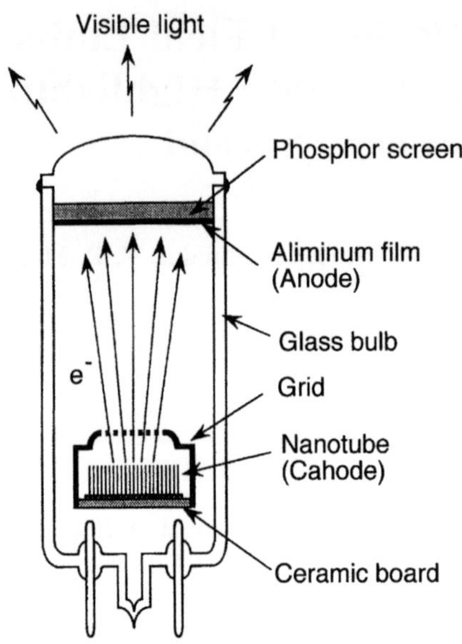

A good understanding of the field emission properties of CNTs is paramount to improve these devices and reduce the failure mechanisms. Therefore, the purpose of my work is a fundamental investigation of the field emission of CNTs.

The understanding of the emission mechanism of CNTs also allows us to use them in other applications that will be listed below. I shall pay special attention of using CNTs as a source of electrons in a Retarding Field Analyzer (RFA). The first scientific papers on CNTs properties reported extremely low turn-on fields, high current densities, good field emission stability compared to metallic emitters in various devices since 1995 [8]; however, the energy distribution of the emitted electrons was not particularly well understood, CNT sources continues offer several attractive characteristics such as instantaneous response to electric field variation, resistance to temperature fluctuation, and high degree of focusability in electron optics due to their sharp (0.2–0.3 eV) energy spread [57, 60]. An RFA is pre-eminently suited to study the energy distribution and the low-energy of an electron source [88, 16]. The first idea is characterize the CNT source and then use the RFA system in the work function measurements of materials.

The main applications of CNT's are:

a) Lighting elements [50, 71], i.e., produce light by bombarding a phosphor-coated surface with electrons. The first device with a CNT cathode to be demonstrated was the field-emission lamp, Fig. 1 [11, 71];

b) Over-voltage protection with nanotubes, in this case the over voltage between a nanotube cathode and a counter-electrode reaches a threshold value for field emission, the emitted current induces a discharge in the noble gas [67];

c) Flat-panel field-emission display [14, 71], the CNT electron source provides a high-brightness display. Field-emission panel displays were demonstrated 10 years ago [98] using Spindt-type emitters [75]. An inherent problem with FEDs is the need for vacuum between the anode and the cathode. Degradation of the vacuum results in ionization of the residual gas by the emitted electrons and poisoning of emitting material resulting in degraded performance. The difficulty in obtaining a robust well packaged display with a long lifetime maybe one reason why the FED has not succeeded commercially. However the main issue is almost certainly the success of the AMLCD. The improvements in LCD quality have reduced the potential market for FEDs to such an extent as to make them commercially unattractive [68];

d) A FED-based backlight unit for LCDs [34, 35] could have lower power consumption than the cold cathode fluorescent lamp and there is severe competition since the development of emitting diode (LED) backlights [68];

e) X-ray sources such as, hand-held X-ray spectrometers and mini-X-ray tubes for medical and other applications [76, 87];

f) High-resolution electron-beam instruments such as electron microscopes, electron-beam-assisted-deposition instruments and electron-beam-lithography instruments [62, 9, 91];

g) Various types of sensors can be applied in different segments of the industries, such as biomedical, automotive, food, agriculture, fishing, manufacturing, security, environmental monitoring and others [69, 93];

h) Transparent conductive thin films for certain niche applications such as organic light emitting diodes (OLED) or Organic photovoltaic (OPV) devices [85, 92];

i) Carbon nanotube electron sources for electron microscopes and other electron beam equipment [60]

j) Power transmission lines with CNTs cables to transport the electricity [2]. The cables exhibit high current-carrying capacity of $10^4 \sim 10^5$ A/cm^2 and can be joined together into arbitrary length and diameter, without degradation of their electrical properties [95];

k) Supercapacitor electrodes prepared from thin films of carbon nanotubes [51, 53];

l) Carbon nanotube transistors are considered the replacement for silicon technology due of their characteristics of low operation voltage. [38].

Carbon nanotubes are one of the most important materials under investigation for nanotechnology, suggesting potential applications in different fields of scientific and engineering, as were described above. The first section of paper is the introduction; the second presents a brief description on basic properties of CNTs. The third section give a review of the electron emission properties by field emission, the fourth section discuss the applications in Field Emission Microscopy (FEM) and fifth section discuss electron Source with a comparison of energy spread between different sources. In sixth section present the measurements of energy-spread, the next sections are conclusions and acknowledgments.

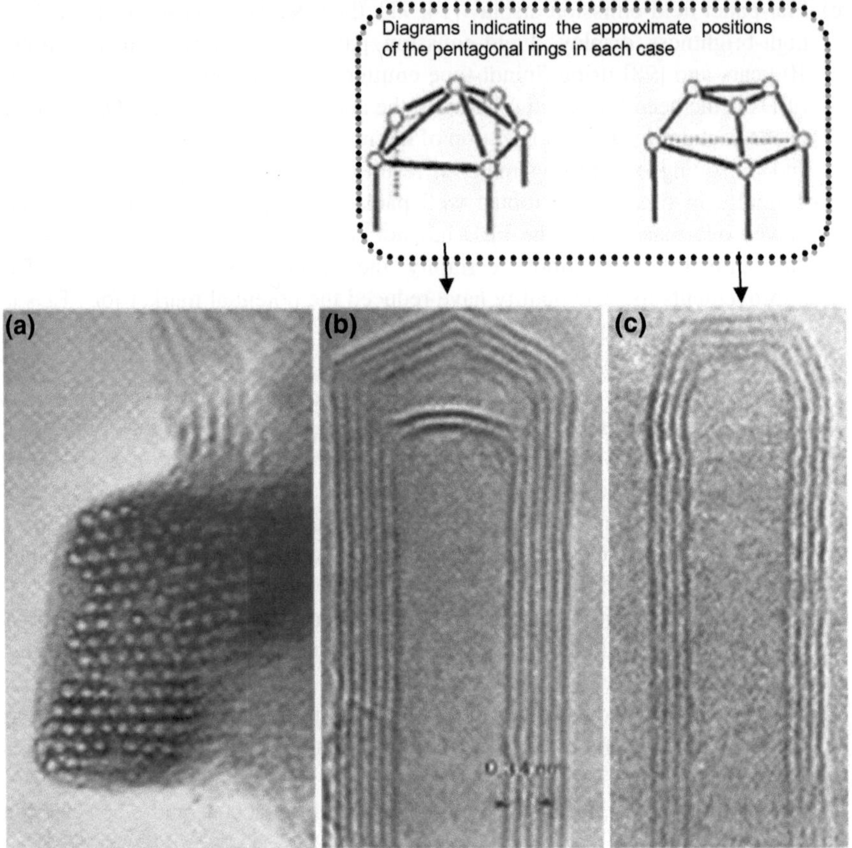

Fig. 2 High resolution transmission electron microscopy (HRTEM) images of CNTs: **a** SWNT micrograph of individual rope, **b** and **c** MWNT Micrograph (by Iijima) with the cap drawing [45]

2 Carbon Nanotube

Carbon nanotubes are formed by one or more graphene sheets rolled to form a cylinder with hollow inside and closed ends. There are two main types of CNTs: Single Wall Nanotube (SWNT), Fig. 2a, consists of one graphene sheet rolled with cylindrical form and the MultiWall Nanotube (MWNT) consists in several concentric graphene tubes with an interlayer spacing of 0.334–0.340 nm, Fig. 2b, [45].

The layer structure of MWNT shows arrangements like "Swiss-roll" and "Russian doll". These two possible arrangements are illustrated in Fig. 3 and also the variation of these two arrangements are a "papier mâché" suggest by [96] and the model by Amelinckx et al. [3] and [45]. The knowledge of these arrangements is important to understand the results obtained by several authors throughout this review.

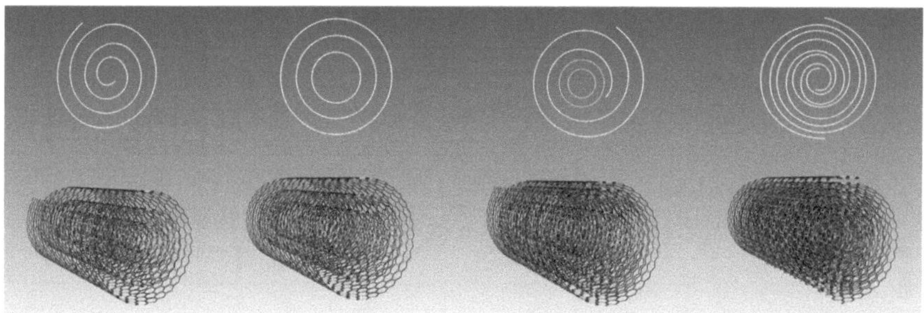

Fig. 3 Schematic drawings (*top* and side view) of arrangements s for MWNT **a** "Russian doll", **b** "Swiss roll" **c** "Papier mâché" suggested by [96] and **d** modeled by Amelinckx et al. (1995)

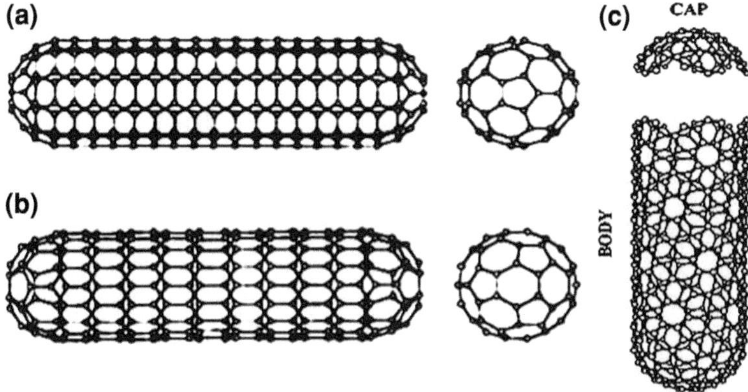

Fig. 4 Schematic drawings of CNTs, example of: **a** armchair, **b** Zig-Zag [45]. and **c** Chiral fiber with hemispherical caps at both ends [69]

CNTs have three basic structures or families. The direction in which the graphene sheet is wrapped is represented by the chiral vector with indices (n,m). The name of these families are Armchair (Fig. 4a) if the chiral angle θ is equal to 30° and the indices n = m $\neq$ 0, Zigzag (Fig. 4b) if $\theta = 0°$ and m = 0 and n $\neq$ 0 and Chiral when the angle $0^0 \leq \theta \leq 30^0$ and n $\neq$ m $\neq$ 0. The lengths of SWNTs and MWNTs are usually well over 1 μm and their diameter range from 1 nm (SWNTs) to 50 nm (MWNTs). Virtually all of the tubes are closed at both ends with caps by fullerene like half spheres, but the tubes can be open at one end or both ends and can be semiconducting or metallic conductors. Observe the example in Fig. 4.

The most common techniques used for CNT synthesis are vaporization methods like arc discharge or laser ablation and the catalytic decomposition of hydrocarbons over metal catalysts or chemical vapor deposition (CVD) [45, 91].

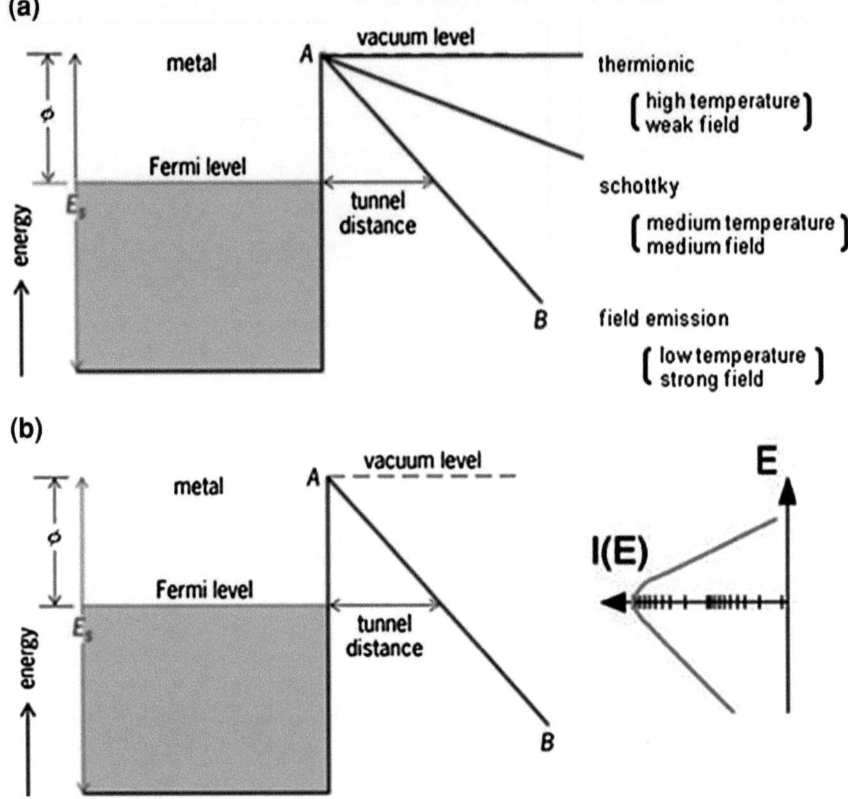

Fig. 5 Diagram of the energy level for electron emission from a metal at absolute zero temperature. **a** comparison between different types of electron emission and **b** diagram of electron field emission

3 Field Emission

The emission of electrons from a metal or semiconductor into vacuum under the influence of a strong electric field is explained in terms quantum-mechanical tunneling. In theory there is a finite probability of the electron being found on the other side of a barrier. Although the electron total energy is lower than the barrier potential. The electrons will be tunneling through the potential barrier [48], rather than escaping over it as in thermionic emission or photoemission.

The process is shown in Fig. 5. The metal has an intrinsic potential and the Fermi level is filled with electrons. The distance between the Fermi level to vacuum level is called the work function (ϕ). The vacuum level represents the potential energy of an electron at rest outside the metal, in the absence of an external field. In the presence of a strong field, the potential barrier will be deformed along the line AB, so that a triangular barrier is formed, through which

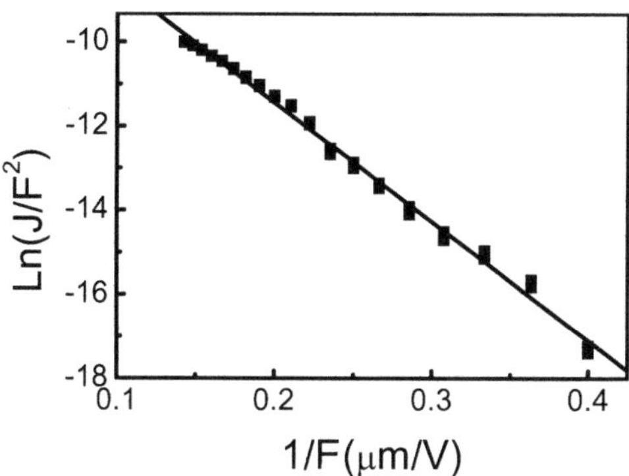

Fig. 6 Fowler-Nordheim plot example of the field emission current density

electrons can tunnel. Most of the emission occurs it the vicinity of the Fermi level where the barrier is thinnest.

In other words the field emission involves the extraction of electrons from a solid by tunneling through the surface potential barrier. The Fowler–Nordheim theory describes the field-emission process. The current density (J) of tunneling is given by Eq. (1):

$$J = \frac{e^3 F^2}{8\pi h \phi t^2(y)} \exp\left\{ -\frac{8\pi\sqrt{2m}\phi^{3/2}}{3heF} v(y) \right\} \tag{1}$$

where, m is the electron mass, F is the electric field, h is the Planck's constant, e is the electron charge and ϕ is a work function. A plot of log (J/F^2) versus 1/F (Fig. 6), the so-called Fowler–Nordheim plot, is approximately a linear curve.

The t(y) and v(y) were calculated by Good and Mueller [24] and can be approximated by $t(y) = 1 + 0.1107y^{1.33}$ and $v(y) = 1 - y^{1.69}$. The y is expressed by Eq. (2)

$$y = \frac{1}{\phi}\sqrt{\frac{e^3 F}{4\pi\varepsilon_0}} \tag{2}$$

ε_0 is the permittivity of free space.

In the case of a triangular surface potential barrier, such that t(y) and v(y) are unity, the current density function by can be approximated as shown Eq. (3)

$$J = 1.54 \times 10^{-6} \frac{F^2}{\phi/e} \exp\left\{ -6.83 \times 10^9 \frac{(\phi/e)^{3/2}}{F} \right\} \tag{3}$$

Here ϕ/e is the work function in electron volts [23, 24]. The model is valid for emission from flat surfaces at 0 K, but it was adapted to describe field emission from sharp tips up to temperatures of several hundred degrees Celsius. Corrections of this model are required for tips with extremely curved surfaces. An additional correction may be necessary in the case of nanotubes since the density of states is not energy independent around the Fermi level as in 'real' metals [70].

The Fowler–Nordheim model shows the dependence of the emitted current on the electric field and the work function. The intensity of electric field shows dependence with some variables as shape of tip, voltage applied and distance between cathode and anode. As a consequence, a small variation of the shape or surrounding of the emitter (geometric field enhancement) and/or the chemical state of the surface has a strong impact on the emitted current.

The local field enhancement factor β is often introduced in the Fowler–Nordheim equation to represent the geometrical effects at the surface of the cathode, where $\beta = F_n/F_0$ for macroscopic applied field F_0. Local variation of β determines the local normal surface electric field, F_n, resulting in local dependence of injection current by the Fowler–Nordheim law. A constant macroscopic electric field $F_0 = 1$ U/m is applied setting the top boundary at a constant surface charge density given by $\sigma = \varepsilon_0 F_0$, where ε_0 is the permittivity of vacuum. In most field enhancement simulation studies the top boundary is at constant voltage U_{anode} in which case the evaluation of β depends on the anode-cathode distance. The maximum field, F_{max}, normalized by the F_0 is the field enhancement factor ($\beta = F_{max}/F_0$). For capped CNTs with radius R, hight h and separation L the aspect ratio $f = h/R$ and normalized separation $s = L/h$ are sufficient to determine β.

These models are applied to a hemisphere on a cylinder, which corresponds to a nanotube with a smooth and clean hemispherical cap. In real situations, however, this condition is not often met. The tip of nanotube emitters (especially for MWNTs) is seldom hemispherical but tapered, and may furthermore be flat, opened and/or present nanometer sized protrusions. Numerical simulations reveal that a deviation from the hemispherical shape produces an increase in the field-enhancement factor up to a factor of two, which can easily lead to a factor of ten in the emitted current.

Irregularities on an atomic level of the tube end are not expected to influence the emission properties for a metallic cap, as the electron cloud smoothes out these irregularities as in metals [41].

3.1 Field Emission Energy Distribution

The Density Of States (DOS) of a system describes the number of states per interval of energy at each energy level that are available to be occupied. In metals, the DOS of the conduction electrons, which are responsible for the emission, is described by the Fermi–Dirac statistics. Above the Fermi level the tunneling probability increases, but the DOS decreases very sharply. Below the Fermi level the DOS increases slightly but the tunneling probability decreases strongly.

These considerations are directly reflected in the specific shape for the Field Emission Energy Distribution (FEED) of the electrons predicted by the Fowler–Nordheim theory. The FEED peaks around the Fermi level with exponential tails that depend on the Fermi temperature of the electrons and on the slope of the tunneling barrier for the high and low energy tail, respectively [36, 40]. Any

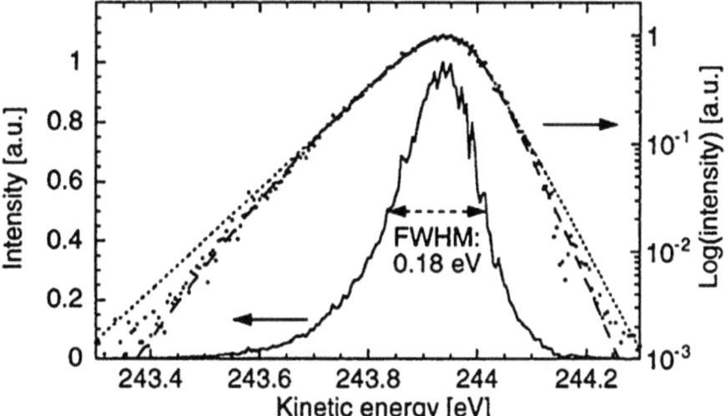

Fig. 7 Field electron energy spectra obtained on a MWNT film showing a single peak, along fits obtained with the F–N distribution (*dotted line*) and with the modified F–N distribution including a Gaussian band of states (*dashed line*) [6]

deviation from this metallic shape is due either to adsorbates or to a nonmetallic DOS. FEED can therefore be used to gain information on the DOS of the emitting electrons as well as to determine ϕ. The Full Width at Half Maximum (FWHM) of the FEED is an expression of the extent of a function, given by the difference between the two extreme values of the independent variable at which the dependent variable is equal to half of its maximum value, for example, in the metal is typically 0.45 eV [40]. Figure 7 shows a typical energy distribution obtained from a film of MWNTs just at the onset of emission. The FWHM is in this case 0.18 eV only, and it was reported an average FWHM over 10 samples of 0.2 eV, without taking into account the broadening due to the finite resolution of the energy analyzer [6]. The experimental setup to Fig. 7 was 3 mm diameter cylindrical counter electrode was placed at a distance of 125 μm for the film emitters and the measurements were carried out at pressures of 10^{-7} mbar.

The shape of the FEED of MWNTs therefore strongly suggests that the electrons are not emitted from a metallic continuum, but from energy bands of 0.2–0.4 eV width. Observe in Fig. 8 the models for the emission.

Fransen et al. [39] observed two kinds of behavior with the applied field for single MWNT emitters. Some spectra showed one peak of ~ 0.3 eV FWHM that shifted with the field and others with peak of ~ 0.15 eV FWHM did not shift. The small FWHM was attributed to the presence of resonant states in the DOS at the tube cap.

Dean et al. [29] detected one peak located at the Fermi level on a single SWNT at room temperature consistent with a metallic, localized or adsorbate state at the Fermi level, Fig. 8b). No peak shift was observed on changing the applied field. A decrease in the emitted current was attributed to adsorbate removal. Since the FEED is a convolution of the tunneling barrier and the electronic DOS of the nanotube, they concluded that tunneling states are present above the Fermi level in clean SWNTs, Fig. 8a).

The low energy showed in FEED spectra from individual SWNT has unusual features at room temperature [58] and were attributed to singularities in the DOS of the SWNT [69, 83], that is, some features of the DOS of the tubes or caps, such as resonant localized states and/or singularities, are reflected in the FEED spectra [58]

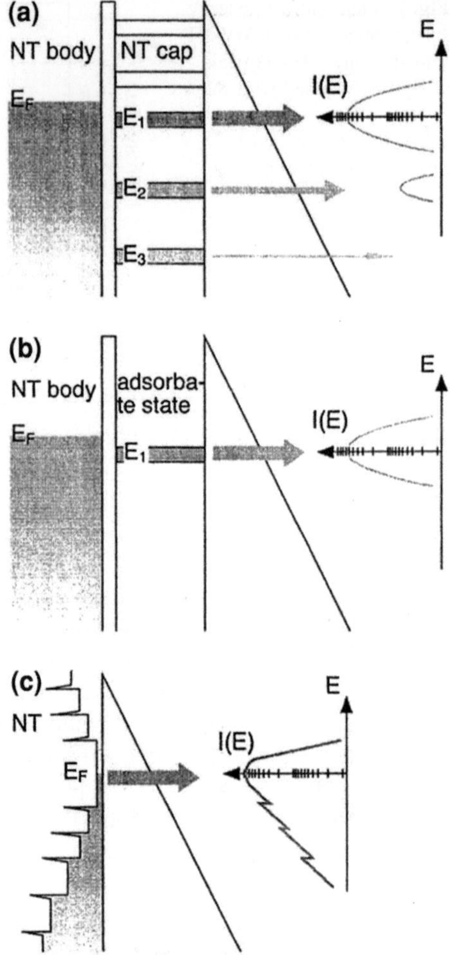

Fig. 8 Models for the field emission showing the energy bands, and the corresponding FEED: **a** emission through energy bands corresponding to DOS at the nanotube cap, **b** adsorbate resonant tunneling, **c** emission from a typical metallic SWNT DOS [8]

Experimental results from [39, 55, 58] on the work function are not conclusive. Fransen et al. determined a workfunction of 7.3 ± 0.7 eV on one MWNT. Kuttel et al. found a workfunction in the range of **5** eV for a CVD MWNT film which was refined to **5.3** eV in a subsequent study. Lovall et al. deduced a value of **5.1** eV for a SWNT. It is not clear if the work functions of closed MWNTs, open MWNTs or SWNTs are different.

It seems far safer to make an assumption on the work function and to deduce the field amplification and emitter shape than work the other way around.

FEED is a reliable method to determine simultaneously the work function of a field emitter and the field amplification. Another observation by ultraviolet photoelectron spectroscopy measurements performed on MWNT films gave a clear indication that the work function can vary significantly with the surface state of the

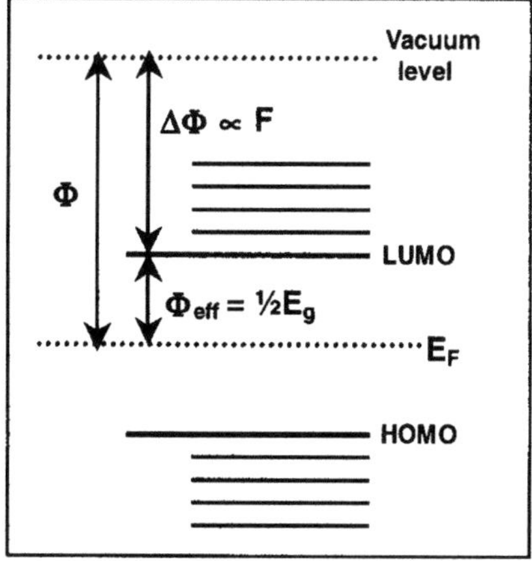

Fig. 9 Schematic drawing showing the effective work function (Φ_{eff}) with the applied electric field [54]

tubes [1]; in other words, the properties may be significantly influenced by the synthesis and purification methods [8].

3.2 Charged States

Several groups reported that the local DOS at the tip showed sharp localized states that were correlated with the presence of defects and also with various tip geometries. The DOS, in the Fig. 9, shows the effective work function, ϕ_{eff}, which is defined as the difference of the energy between the Fermi level and the lowest unoccupied molecular orbital level (LUMO), i.e., half the band gap in case of a semiconductor. The difference between the LUMO and the vacuum level $\Delta\phi$ would vary when the strength of the applied electric field is changed. The effective work function of the capped armchair nanotubes decreases linearly with increasing electric fields, whereas that of the metal tip decreases quadratically. The Mulliken charge population shows that the charge accumulation is not dependent on the local atomic geometry but on the sharpness of the tip [54].

Figure 10 shows change of highest occupied molecular orbital (HOMO) and LUMO for various charged states under the applied electric field for neutral states, $Q = -2e$ and $Q = 2e$. Drawings showed in the left are local charge densities under no applied electric field and the right column shows those under an electric field of 1 V/Å. It is possible to observe where the charge transfers occur. For negatively charged states, the HOMO and LUMO are localized at the cap under an electric field. Because the local field can be enhanced easily at the cap by the local field enhancement factor, these electrons can be easily emitted by the external electric field. For positively charged states, on the other hand, the HOMO and LUMO are not always localized at the cap under an electric field. These states are not affected much by the external electric field.

Fig. 10 Charged states drawings of armchair nanotube cap (5,5). The left column shows local charge densities under no applied electric field and the right column shows those under an electric field of 1 V/Å.
a Neutral state ($Q = 0$);
b $Q = -2e$ and **c** $Q = 2e$ [54]

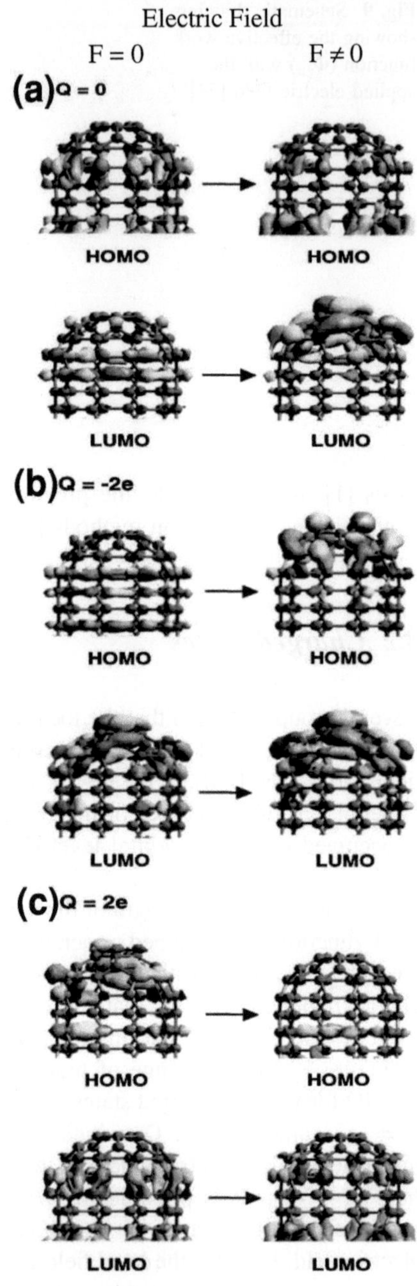

3.3 Field Emission Mechanism

Rinzler et al. [66] proposed a model that higher emission efficiency is from open MWNTs but this could not be confirmed experimentally yet, and probably will be valid only for open tubes [64, 63]. Obraztsov et al. [63] proposed that model would be valid only for open MWNT.

The observations of the experiments is possible to consider that the electrons are emitted from sharp energy levels due to localized states at the tube cap [26] like in the theoretical calculations and was confirmed experimentally by scanning tunneling microscope (STM) measurements on MWNTs [10], as well as, on SWNTs. The FWHM of these states and their separation is in good agreement with the values measured for the light emission and with FEED observations carried out at 20 °C and at 600 °C [30]. In this hypothesis is necessary to consider two points. First, if several energy levels participate in the emission, the occupied level nearest to the Fermi energy will supply nearly all the emitted electrons and this level is strongly depend of atomic configuration. Therefore, exist significant differences of the emitted currents from one tube to another and second it is necessary observing the carrier densities in the states because the field emission current depends directly on this carrier density [8].

Dean et al. [29, 31] suggest that nanotubes emission is more complex than the metallic tip emission. Different emission regimes on single SWNTs were identified and depend on applied field and temperature. One regime is a resonant tunneling through an adsorbate, in this case, the water molecule. These molecules desorb either at high fields and emitted currents or at temperatures higher than 400 °C. The other regime correspond to the intrinsic emission from the cleaned tube and show a far lower emitted current for comparable voltages with strongly reduced current fluctuations. The origin of these intrinsic regimes is not clear yet, but the emission mechanism involves probably non-metallic electronic states, such as enhanced field emission states above the Fermi level or a non-metallic DOS [8].

Thus, the results observed by different groups shows that emissions involve a non-metallic DOS and/or adsorbate-resonant tunneling. Supplementary information on electronic and structural properties of individual nanotubes, on the nanotube cap and on the influence of absorbates or bonded groups are required for a better comprehension of the emission.

3.4 Setup for Arrangements of CNTs Emissions Experiments

In the electron emission of the CNTs normally it is necessary to use a support that in some cases is a substrate to CNTs films, a wire or a filament to individual nanotube. Following is a description of some types of setup.

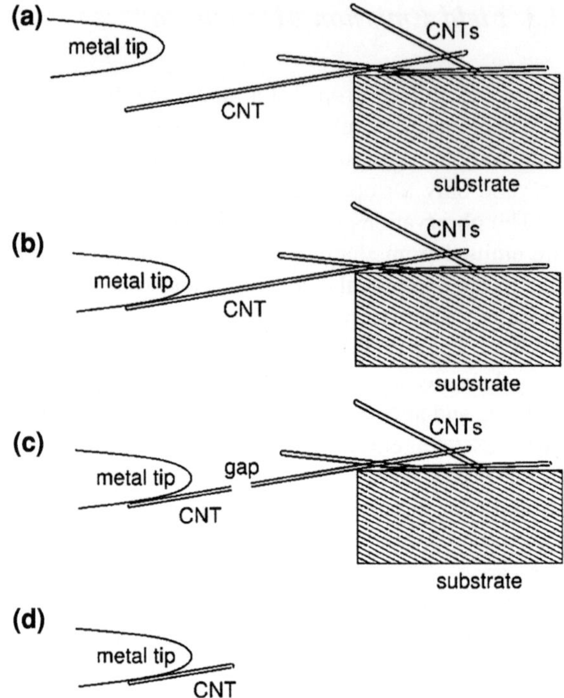

Fig. 11 **a** A CNT protruding from a thin substrate containing many nanotubes is selected and **b** attached to a tungsten support tip. **c** The CNT is broken by Joule heating. **d** The open tube end is finally closed [25]

a) The nanotubes are mounted on a support tip by a simple method. Under an optical microscope equipped with two three-axes micromanipulators used to move independently two supports, normally one to the tungsten filament and other to move nanotubes with conductive characteristics [66, 39]. Individual carbon nanotubes stick to the tip either by Van der Waals forces alone or by first applying a bit of adhesive to the tip. The resolution of an optical microscope is not sufficient to observe one nanotube and it is therefore necessary to characterize the emitters by scanning or transmission electron microscopy [4].

b) To improve the previous method it is necessary to use a piezo-driven nanomanipulator in a Scanning Electron Microscope (SEM). A tungsten wire is fixed by laser-welding on a filament. The tip is transferred into the SEM and carefully pierced into carbon tape. In this case also it is used a glue on the tip for firm attachment of the nanotube [22]. These nanotubes were broken by running a current through it the break-off occurring at a weak spots along the length of the nanotube (Fig. 11c), possibly a defect in its structure. Figure 11 shows the schematic drawing of the procedure. [25].

c) Fix the individual nanotubes on the support, for instance, using low-energy electron microscope equipped with micromanipulators, or using a mat of SWNT material in this case after the contact with the support under the optical microscope, a small voltage (10 V) is applied to break the attached rope from the mat, another method operates inside a scanning electron microscope [58]. Individual nanotubes can be picked up and attached to AFM cantilevers by electron beam irradiation [21]. These deposits are mechanically strong and the electrical contact between the tube and the support should be strongly enhanced by the electron-beam irradiation and finally the direct growth of one nanotube on a support by CVD [15, 84].

d) In the case of carbon nanotubes films can be made from 'bulk' samples (by pressing nanotube powder or spraying a nanotube-containing suspension on a support) or can be grown directly on desired positions on a support structure using CVD techniques. However, the relation between the morphology and the emission properties cannot be properly investigated, since the emission of a single nanotube is different from the emission of a nanotubes film [9].

It was observed in some experiments with impure material it is more difficult to isolate an individual nanotube, in these cases a macroscopic fragment is picked up and is fixed on top of a filament tip. Purified material is better; the tubes are segmented and well separated in the solution following an oxidation treatment. But the effect of the purification of the tubes is not well known and the cap removal is not controlled and may affect the emission properties.

3.5 Field Emission from CNT Films

After Iijima identified the carbon nanotubes in 1991, [12, 49] reported field emission from "tubulene" films (MWCNT). Due to its high density, the tube protruded only a few nanometers above the surface and consequently the voltages needed to extract the current were very high (<25 V/μm). De Heer et al. [19] observed electron emission from a continuous film of randomly oriented (MWNT), with macroscopic current densities as high as 100 mA/cm^2.

The field emission is excellent for nearly all types of nanotubes. The threshold fields are as low as 1 V/μm (minimum field to emission) and turn-on fields are around 5 V/μm (ideal field to emission). Nanotube films are capable of emitting current densities up to a few A/cm^2 at fields below 10 V/μm. One interesting parameter is the emitter density on the films [8]. Typically, a film has a nanotube density of $10^8 \pm 10^9$ cm^{-2}. The effective number of emitting sites, however, is quite low. Typical densities of $10^3 \pm 10^4$ emitters/cm^2 were reported at the onset of emission [19, 5, 97]. By using an optical microscope combined with a phosphor screen, [63] was able to enhance the resolution of the measurement and reported densities of $10^7 \pm 10^8$ cm^{-2} [8].

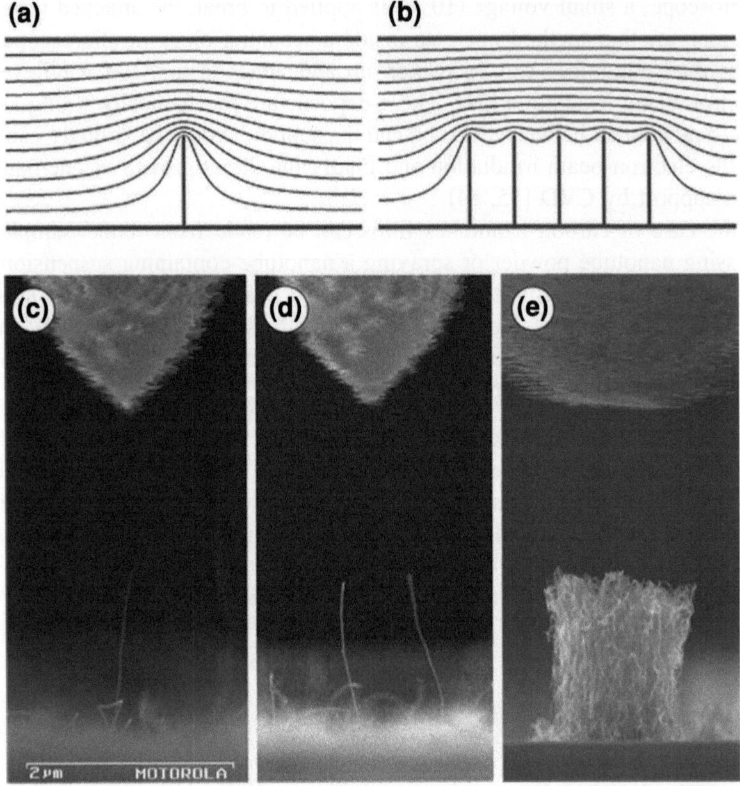

Fig. 12 **a** Equipotential lines of the electrostatic field for an single CNT and **b** with five CNTs showing the electrostatic screening. The others are images of three experimental situations by SEM; **c** Emission was observed at 90 V, **d** and **e** Emissions were observed up to 200 V in both cases, the results were the same [23]

3.5.1 Emission Characteristics

Electrostatic Screening

Several studies reveal that the electric field at the apex of the emitters decreases with decreasing spacing between the emitters, more exactly for a spacing less than twice the height [7, 61].

In other words, the field enhancement is largest for a single nanotube and decreases as the as nanotubes are positioned close, as shown in Fig. 12. [43, 61]. So, the optimum configuration is the largest distance with the highest density.

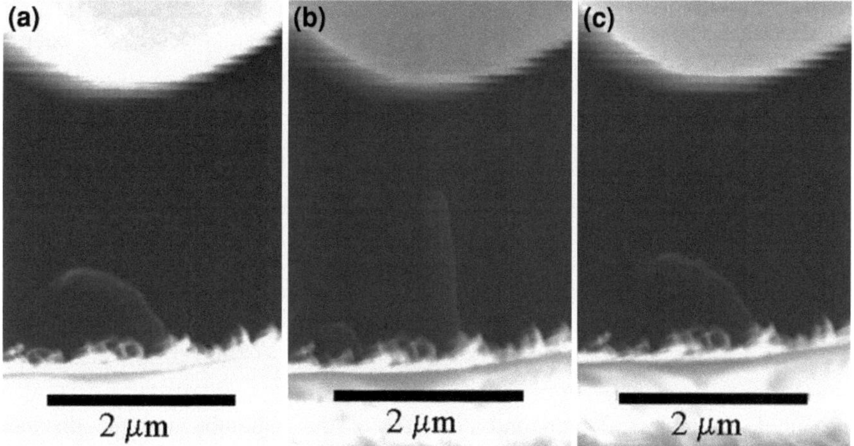

Fig. 13 SEM images showing the CNT bending under different electric fields **a** 0 V, **b** 20 V, and **c** 0 V. Observe the alignment of CNT with the electric field lines [82]

Height and Diameter of CNT

Theoretical considerations on the field emission of CNTs indicate their outstanding feature, because of their high aspect ratio. That is the length of a vertically oriented CNT divided by the diameter. From this point of view long vertically aligned CNTs should be super cold emitters.

But experimental results showed that less densely populated "short and stubby" nanotubes showed the best emission characteristics with a threshold voltage of 2 V/μm and saturation emission current density of 10 mA/cm^2 [13].

Alignment in an Electric Field

The influence of an applied electric field on CNT was observed, CNTs flexed to orient themselves parallel to the electric field lines. For moderate field strengths below the electron field emission threshold, the flexed nanotubes relaxed back to their original shapes after the electric field was removed, Fig. 13 [82].

Emitted Current

The emitted current varies in an exponential way with the electric field. A small variation in the field produces a huge difference in the emitted current (this is demonstrated by the Fowler–Nordheim model). The consequence is difficult to predict the emission with precision, for instance, in the experiments measurements of some individual nanotubes to compared with the values predicted from the

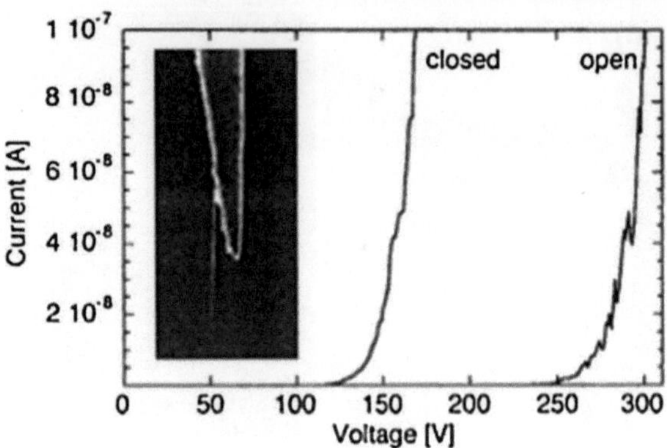

Fig. 14 Field emission I × V Curves acquired on a cap of the SWNT and MWNT on etched gold fiber [8]

measured parameters; length and diameter of the nanotube and the distance between the cathode and anode resulted in two very different values. These results are very interesting because is possible to conclude that the emission process is highly sensitive to the exact structure of the cap. Other possible conclusion is that for a film just a single or a few nanotubes are necessary to provide the emitted current for a pixel area because a nanotube needs to be only slightly thinner, longer or sharper than the surrounding nanotubes to dominate the emission. This effect can even be increased when nanotubes have an adsorbed species on the cap.

This effect in the emitted current has been investigated by comparing the measurements on a nanotube film and on individual emitters from the same film and this experiment showed that the field enhancements are far higher in the former case (nanotube film) than in the latter (individual nanotube) [9]. This effect is critical for most devices, where an homogeneous emission and high emission site density is needed. In fact, the dispersion on γ (and hence on the combination of nanotube height, length and spacing) should be less than 4 %, which is difficult to achieve even when the growth is well controlled [73]. One possibility to avoid this problem is to include a suitably scaled ballast resistor in series with the emitter to produce a voltage drop.

3.6 Emission from Individual CNTs

The first electron field emission from a single nanotube was reported by [66]. He studied MWNT mounted on a carbon fiber with emission currents of 100 nA at 0.12 V/μm. [8] studied closed and open MWNT nanotubes mounted on gold fibers. It was found that open tubes emitted at about twice the voltage needed for the closed ones, as is shown in Fig. 14. In contradiction Saito et al. found that open MWNTs began to emit electrons at the lowest fields. The results are inconclusive so more research is needed in this area [8].

3.6.1 Emission Stability

The Fowler–Nordheim model is used to explain the tunneling current in field emission although the I x V characteristics of nanotubes do not follow the predicted behavior over the whole current range model [26, 4]. Two different current regimes were observed: fluctuations at low emitted currents with a switching frequency that increased with the current and became maximal at the onset saturation, followed at higher currents by stable emission with flicker noise [71]. Observed current fluctuations at low currents and attributed the observed saturation to the presence of non-metallic resonant states at the cap.

Stable emissions were observed by several groups [19, 37, 90, 59], the highlight is [71]. They observed the emission during 8000 h with emission current of 10 mA/cm^2. The problem of the instable emission can be due the degradation and can be reversible or permanent, but the origin is not clear. The other possibilities observed for the instability are the residual gases and the intrinsic proprieties of nanotubes. Comparing films of SWNTs and MWNTs in the same chamber pressure conditions and emitted current density observed a faster degradation of SWNT probably because the single wall is more sensitive to ion bombardment and irradiation [26, 5].

Failure

Several aspects that provide failure were observed. The electrostatic deflection or mechanical stresses can cause alterations in the shape and/or surroundings of the emitter, which may lead to a decrease in the local field amplification. High currents can rapidly damage a nanotube. A gradual decrease in field enhancement due to field evaporation was found on SWNTs when the emitted current was increased beyond a given limit between 300 nA–1 μA [32].

On MWNTs, a shortening of the emitter over time, (Fig. 15) [82] or damage to the outer walls of the nanotube due to high currents [81] were also reported. The CNTs can be operated at currents of up to 10 μA, but the current should be kept below 1 μA for stable operation and long life-time [20].

Bonard et al. [8] observed reversible degradation in the emissions of CNTs films and permanent for the individual nanotube is in nearly all cases abrupt, the emission happened on most emitters as a irreversible failure that occurred in less than 10 ms.

De Heer et al. [19] presented recently some experiments of field emission on single MWNTs in a transmission electron microscope. It appears that tube failure occurs on a very short time scale (<1 ms) at currents above 0.1 mA and that it involves an irreversible damage to the tube. Tube layers or caps are removed, peeled back, or the end of the tube is amorphized [81]. In all cases a strong decrease in the emission current occurs and the voltage has to be substantially increased to obtain comparable currents. Cumings et al. [18, 17] observed single MWNTs with a similar setup. They were able to "peel" the tube layer by layer by

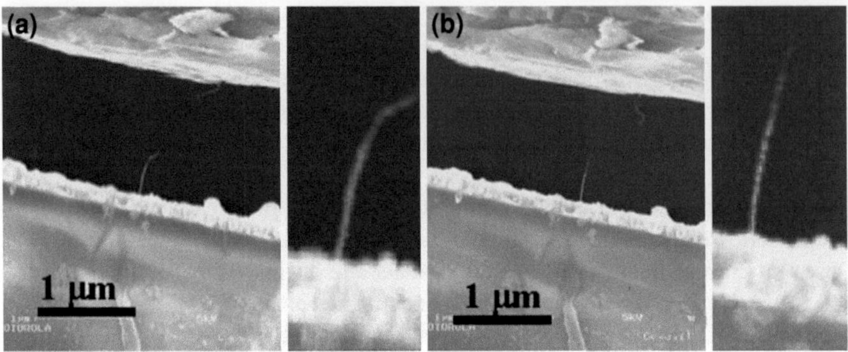

Fig. 15 Images by SEM showing the permanent deformation of the CNT after field emission: **a** image before field emission; **b** Image obtained after field emission and after the strong electric field was removed. The nanotube stayed in the straight shape [82]

applying a strong current through the tube. [26] proved that the location of the electrical failure on a single MWNT could be correlated with the presence of a defect in the nanotube.

Another problem is arcing, i.e. an arc between cathode and anode that is initiated by field emission. Such arcing events are commonly observed on diamond and diamond-like carbon films [45]. They are usually caused by a high field-emission current, anode out-gassing, or local evaporation of cathode material that creates a conducting channel between the electrodes.

Fluctuations

The fluctuations of the emitted current over time are caused by the exponential behaviour of field emission: small changes in the electric field and the work-function have already large effects. Small molecules absorbed on the cap may lead to slight changes in the workfunction or of the cap geometry, resulting in large differences in the emitted current, in other words, the nanotube emitter is sensitive to surface diffusion of emitter material in the presence of the large electric field required for field emission [44]. Dean et al. [31] observed this phenomenon on single SWNTs. These fluctuations disappeared nearly completely at higher currents. This behavior was ascribed to the presence of adsorbates (most probably water) at the cap that enhance the field emission as compared to clean caps [23].

There are several ways to reduce the emission current fluctuations.

1. Cleaning of the CNTs, but the main problem is that the clean nanotube is sensitive to molecules in the vacuum, and especially to oxygen, which can cause irreversible damage to the tube [27].
2. Heating the emitter to 1000 K in ultra-high vacuum to remove adsorbed molecules or impurities [27, 22, 70, 86]
3. Self-heating of the nanotube to remove adsorbed molecules [65]

4. Using a series resistor of several $M\Omega$ to regulation of the current (ballast resistor)

5. The stability can be largely increased when keeping the temperature at 600–900 K, so that absorbed molecules are continuously driven off the nanotube [20]

Noble gases and H_2 usually cause an increase in the current fluctuations, but the emission stability is restored after evacuation of the gas. No degradation was measured on a single nanotube after exposure to Ar and H_2. Conversely, irreversible continuous decreases in the current were provoked by exposure to oxygen and water and were attributed to reactive sputter etching [28].

The vacuum level should be of the order of 10^{-9}–10^{-10} mbar for emission stability of a few percent, but lower vacuum levels are also allowed at the cost of stability. At lower vacuum levels it would be advantageous to reduce the partial pressure of oxygen and water.

An important question is how stable the average electric field around the cap is in time. It was found by electron holography experiments that the electric field remained remarkably constant, although large fluctuations of the emitted current did take place [18].

4 Field Emission Microscopy

The electron beam that is emitted from a nanotube is not uniform, but shows certain patterns when projected on a phosphor screen. The images projected depend on the type of cap. Emission patterns with a clearly visible symmetry were observed for MWNTs [72], and after heated at 1300 K some bright spots showed in the image disappeared in time, this behavior was typical for absorbed species on the cap.

The emitted current decreased as each molecule (bright spot) was removed. The current emitted by the clean tube was lower by a factor of three than the "dirty" tube (e.g. gas molecules on the cap): this shows that, on CNTs, adsorbed species enhance the field emission, observe some images in Fig. 16 [31, 46].

SWNT films showed ring patterns that were interpreted as coming from nanotube caps because of the circular symmetry [97]. Others patterns like circular arcs and rings on films were attributed to emission from the open edges [59].

Movements in the patterns were observed. These patterns started to rotate when the applied electric field was above a certain threshold value [33]. They observed similar patterns for samples made of individual multi-walled CNTs mounted on tungsten tips [20].

Bonard et al. [8] also detected bright spots from MWNT and SWNT emitters along with well-defined patterns of two or fourfold symmetry. These patterns persisted after applying high positive fields in desorbing possible adsorbates and reflect probably the electronic density of the emitting states at the nanotube cap.

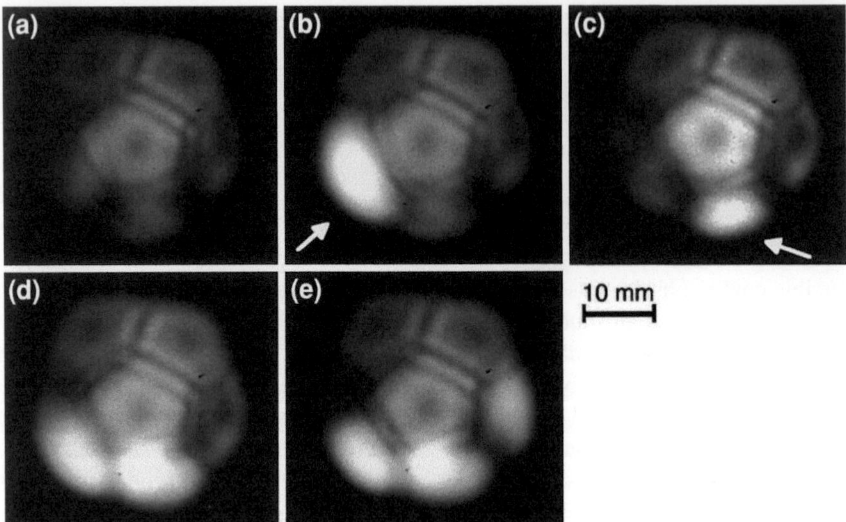

Fig. 16 Images of capped MWNT by FEM; **a** After heat cleaning and **b–e** showing a sequence of adsorption of gas molecules [46]

The non-homogenous density would identify them as localized states at the cap and not as delocalized conduction-band states as in metals.

Dean et al. [29] showed patterns of SWNTs at room temperature with one to four lobes that are typical and behaved like molecules adsorbed on the tube cap. These patterns disappeared above 600 °C, and fine structured patterns with sometimes five or six fold symmetry were detected. Dean et al. [29] argue that these patterns represent the electronic distribution of the emitting states at the nanotube cap and not the atomic structure of the cap.

In some results of field emission microscopy it is possible observe some characteristics such as:

a) a pattern emission with a clearly visible symmetry, but when the symmetry of the pattern is disturbed some authors attribute the problems in the cap shape, because emission process is highly sensitive to the exact structure of the cap, as described in the emitted current section.
b) bright spots, usually explained by the existence of adsorbed species [29] in the emission, but it should also be noted that the cap may contain small protrusions [56].
c) no pattern emission, it may mean problems in the cap. The open caps should produce irregular and changing patterns, due to their sharp edges and dangling bonds. Irregular patterns were also observed for damaged caps, consisting of amorphous carbon.

The images in Fig. 17 exemplify the FEM emissions, the images between a) and f) show the FEM patterns of nanotubes. These patterns are interpreted as the

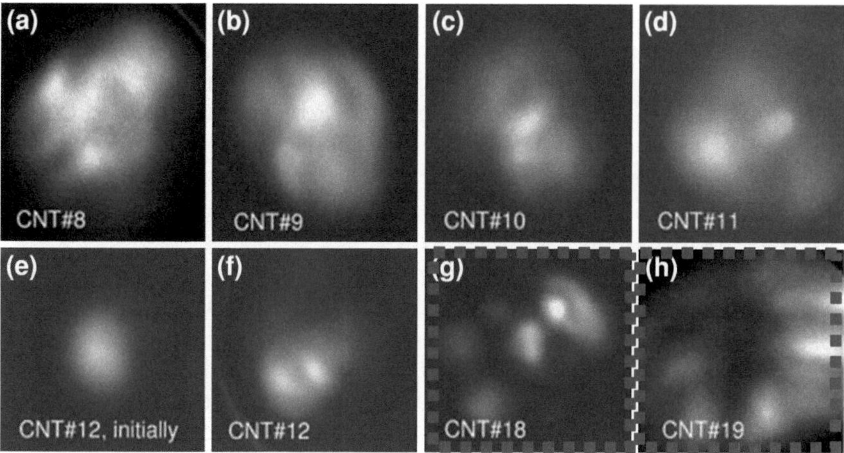

Fig. 17 FEM images of electron sources. The images: **a–f** showing patterns of closed caps; **g** and **h** The emission patterns of CNTs with open caps [25]

typical emission of SWNT or thin MWNT with caps. These emissions were highly stable with time for emitted currents up to 1 μA. The images g) and h) with red dotted lines showing images changing their positions and intensities every few seconds and even rotated and are interpreted as images of nanotubes without cap. The electron emission from a capped CNT exhibits a high current stability in contrast to emission from CNTs with open cap [25].

In summary, some tubes produce homogenous spot or ring patterns that are quite similar to those observed on metal tips. On the other hand, the observation of fine-structured stable patterns strongly suggests that the electronic distribution of the emitting states is not homogenous. The origin of this discrepancy is explained by different cap geometries because the topology of the cap affects the strength and position of the peaks observed in density of states spectra near the Fermi energy. In other words the cap has different electronic structures [52]. Some experiments using TEM, shown that the work function is sensitive to atomic structure and the surface specification of the emitting tips [52].

5 Electron Source

The traditional types of electron sources are the thermionic, Schottky and cold field emission source. The diagram of the electron sources was presented in the first section in Fig. 5, the energy level scheme to each type of source. In the thermionic electron source, a material is heated to a high temperature and its electrons gain sufficient energy to overcome the material's work function to be emitted. The Schottky emitter, sometimes also referred to as the field enhanced thermionic emitter or thermal field emitter, combines heat, low work function, and a moderate applied electric field to create a stable electron source. The Cold field emission is obtained under an applied electric field, in this case the CNT embodies a unique combination of properties which make a potentially and extraordinary electron source [79, 60], the advantages over other field-emitting materials including high aspect ratio (hence a high field enhancement and brightness), strong covalent bonding preventing electromigration of surface atoms and a high conductance, high stability, energy spread as little as 0.2 eV and reduced brightnesses of the order of 10^9 A/srm^{2}V [60].

CNTs have some advantages when applied in electron sources of microscopes: a cold electron source, providing a high current density and high brightness; however, a more important consideration is the effect of the field emission mechanism on the energy spread of the electron beam.

The expected energy spread of field emission can be calculated by expressing the current density as function of the energy using the Fowler–Nordheim theory, Eq. (4) [47]:

$$J(E) = \frac{4\pi med}{h^3} \exp\left\{-\left(E - \frac{2v(y)}{3t(y)}\phi\right)\frac{1}{d}\right\}\left(1 + \exp\left(\frac{E}{k_B T}\right)\right)^{-1} \tag{4}$$

k_B is the Boltzmann constant, T is the temperature, the other symbols are defined in Eqs (1) and (2), and the tunneling parameter d is given by Eq. (5).

$$d = \frac{ehF}{4\pi\sqrt{2m\phi}t(y)} \tag{5}$$

The parameter d can be expressed for t = 1 as Eq. (6).

$$d = 9.76 \times 10^{-11} \frac{eF}{(\phi/e)^{1/2}} \tag{6}$$

The energy spread ΔE of the source is defined by the full width at half maximum (FWHM) of the energy spectrum and is determined by d and T [23].

Several experiments revealed a small energy spread of 0.2–0.3 eV as expected for a cold field emitter and only small deviations from the Fowler–Nordheim theory [6, 22, 43], but in some cases turned out to be somewhat larger of up to 0.5 eV. Data of [46] obtained values between 0.3 and 0.4 eV, and their energy spectra contained a relatively large shoulder at the high-energy side [78]. Probably by self-heating of the nanotube, because the experiments of Purcell et al. [65], in which the effect of self-heating was investigated, demonstrated a significant broadening of the high-energy side of the energy spectra as function of the emitted current.

In some applications such as electron probes for microscopy or lithography or electron beam equipment the stability, energy spread, noise, emission pattern, brightness and current per emitter are important parameters of electron source, as an example the narrow energy spread is desirable as it improves the focus ability of the beam using electrostatic lenses [79, 60]. The carbon nanotubes have some

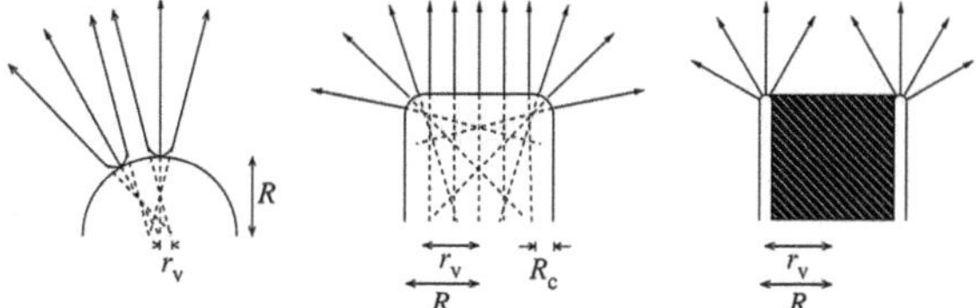

Fig. 18 Electron trajectories **a** Nanotube with Spherical cap and radius R; **b** Nanotube with a flat cap and **c** Nanotube with an open cap [23]

properties that make them attractive material for field emission source as described in the beginning of section.

To perform the energy measurements we should take account the brightness parameter of electron source. Generally there is a direct proportionality between useful beam current and brightness, observe the explanation below.

5.1 Brightness of CNT Electron Sources

The reduced brightness (Br) is another important parameter of electron sources, because it indicates the amount of current that can be focused into a spot of a certain size and from a certain solid angle. Note the Eq. (7), it is a function of the radius of the virtual source r_v, the angular current density I' corresponding to the brightest fraction of the emitted electron beam and the beam potential U [44, 47]:

$$B_r = \frac{I'}{\pi r_v^2 U} \tag{7}$$

The virtual source of an electron emitter is the area from which the electrons appear to originate inside the CNTs [44]. Figure 18 shows some electron trajectories, from which the magnitude of r_v is obtained. The virtual source sizes of individual multi-walled CNTs were determined by operating the nanotubes as electron sources in a point-projection microscope [89]. It was found that the radius of the virtual source size varied between 2.1 and 3.0 nm [20]. This result was larger than expected for the emitter with a hemispherical cap and was explained by a flat, or open cap [23].

The results from reduced angular current density were observed between 19 and 50 nA sr^{-1} V^{-1}. The angular current density was measured for a cleaned nanotube and the maximal current in the stable emission was 1 µA [44, 77, 850].

Table 1 compares the nanotube electron source with the thermionic, Schottky, and metal (tungsten) field emission sources. The tungsten cold field emission source is used in applications where heat is a problem or where a very small energy spread is required. However, it is applied at the expense of poor stability. The Schottky source offers excellent stability at the expense of energy spread and heat.

From Table 1 it can be concluded that a CNT field emitter, which has the same stability and noise as the Schottky source, delivers a high-brightness electron beam with narrow energy spread and therefore, it is a formidable contender as a source for electron optical applications [79].

Table 1 Properties of different electron sources, observe the energy spread to each case [79]

Property	Thermionic (Tungsten/LaB)$_6$	Schottky (Tungsten + ZrO)	Metal cold FE (Tungsten)	Carbon nanotube FE[a]
Virtual source size (nm)	10,000	<20	<10	<10
Energy spread (eV)	1	0.7	0.2–0.3	0.2–0.35
Brightness (A/ m^{2}srV)	10^6–10^7	10^8	10^8	10^9
Stability (%)	<1	<1	4–6	<0.5
Operating temp	1,500–2,100 °C	1,500 °C	25 °C	25 °C–400 °C
Lifetime	100–1,000 h	>1 year	>1 year	>1 year

[a] For the carbon nanotube field emission source, it can either be operated at room temperature (25 °C) or slightly warm (400 °C) to prevent re-adsorbtion of residual molecules in the vaccum and enhance its ability. Even when hot, carbon nanotube,which are covalently bonded, remain stable and do not suffer from diffusion/electromigration like metal emitters [5]

6 Measurement of Energy Spread

The knowledge of the beam energy spread (or width) with accuracy is important because is possible characterize the electron source. A retarding field analyzer (RFA) is our choice to measure the energy spread of electron beams because of its simplicity, compactness and high signal-to noise ratio output.

There are different types of RFAs and the simplest is with parallel plate [16]. It consists of the CNT electron source and two parallel plates. The first plate is grounded and the second one is biased to a negative high voltage to retard the electron beans, only those particles whose longitudinal kinetic energy is higher than the retarding potential can thus pass second electrode and reach the Copper collector plate forming a current signal. Other electrons will be reflected. To improve the system we insulate the mesh electrically from the retarding and apply a controllable small voltage between the second plate and the collector. By varying the voltage the associate electric field provides changes in the slopes of the trajectories so that the beam can be made to pass the lowest potential before the mesh with no transverse velocity [16].

7 Conclusion

The studies of nanotube films with different densities and measured under identical conditions shows that a film of low density and short tubes are inefficient cathode. The medium density films show a very homogeneous and strong emission with a large number of emitting sites. A very dense film, however, shows a decreased quality of the emission. These results from a combination of two effects: the intertube distance and the number of emitters. When the intertube distance is large, the field amplification factor is determined only by the diameter and the height of the nanotube. As the distance between the tubes is decreased, screening effects become significant.

Different properties of carbon nanotubes proved that they are excellent electron sources, providing a stable current at very low fields and capable of operating in moderate vacuum. Different methods are available to deposit of nanotubes on surfaces with different density and orientation and therefore to control their emission properties. The degradation remains also a big unknown in spite of its utmost importance for applications.

Carbon nanotube electron sources clearly have interesting properties, such as low voltage operation, good emission stability, long lifetime, high brightness and low energy spread. Several applications could possibly benefit from the use of CNTs, and considerable research efforts in both academy and industry have been allocated to evaluate these possibilities. The most promising applications are the field-emission display and high-resolution electron-beam instruments. Yet many hurdles remain.

The results of several researchers cannot get a more incisive conclusion because the comparison and interpretation of the results is difficult because most groups use different designs of cathodes, materials and experimental procedures. Also, it is difficult to compare data reported by the various authors since they frequently use different definitions of the emission parameters and their measurements depend upon a variety of experimental conditions, many of which are unreported [94].

These facts make a thorough comparison of results delicate, particularly because the methods used for synthesis, purification, presence of contaminating material, and film deposition are quite varied. The interpretation is further complicated by the different experimental setups, e.g., the uses of planar, spherical or sharp tip anodes, and different distances to emissions.

References

1. Ago, H., Kugler, T., Cacialli, F., Salaneck, W.R., Shaffer, M.S.P., Windle, A.H., Friend, R.H.J.: Work functions and surface functional groups of multiwall carbon nanotubes. Phys Chem B. **103**, 8116–8121 (1999)
2. Alvarenga, J., Jarosz, P.R., Schauerman, C.M., Moses, B.T., Landi, B.J., Cress, C.D., Raffaelle, R.P.: High conductivity carbon nanotube wires from radial densification and ionic doping. Appl. Phys. Lett. **97**, 182106-1–182106-3 (2010)
3. Amelinckx, S., Bernaerts, D., Zhang, X.B., Van Tendeloo, G., Van Landuyt J.: A structure model and growth mechanism for multishell carbon nanotubes. Science. 267, 1334-1338 (1995)
4. Bonard, J.M., Maier, F., Stockli, T., Châtelain, A., De Heer, W.A., Salvetat, J.P., Forró L.: Field emission properties of multiwalled carbon nanotubes. Ultramicroscopy. **73**, 9–18 (1998)
5. Bonard, J.M., Salvetat, J.P., Stöckli, T., De Heer, W.A., Forró, L., Châtelain, A.: Field emission from single-wall carbon nanotube films. Appl. Phys. Lett. **73**, 918–920 (1998b)

6. Bonard, J.M., Salvetat, J.P., Stockli, T., Forro, L., Chatelain, A.: Field emission from carbon nanotubes: perspectives for applications and clues to the emission mechanism. Appl. Phys. **A69**, 245–254 (1999)
7. Bonard, J.M., Weiss, N., Kind, H., Stoeckli, T., Forro, L., Kern, K., Chatelain, A.: Tuning the field emission from carbon nanotube films. Adv. Mater. **13**, 184–188 (2001b)
8. Bonard, J.M., Kind, H., Stockli, T., Nilsson, L.O.: Field emission from carbon nanotubes: the first five years. Solid State Electron. **45**, 893–914 (2001c)
9. Bonard, J.M., Dean, K.A., Coll, F.C., Klinke, C.: Field emission of individual carbon nanotubes in the scanning electron microscope. Phys. Rev. Lett. **89**, 197602 (2002b)
10. Carroll, D.L., Redlich, P., Ajayan, P.M., Charlier, J.C., Blase, X., De, A., Vita, R.: Electronic structure and localized states at carbon nanotube tips. Phys. Rev. Lett. **78**, 2811–2814 (1997)
11. Chen, J., Zhou, X., Deng, S.Z., Xu, N.S.: The application of carbon nanotubes in high-efficiency low power consumption field-emission luminescent tube. Ultramicroscopy **95**, 153–156 (2003)
12. Chernozatonskiib, L.A., Gulyaevb, Yu. V., Kosakovskajab, Z. Ja., Sinitsync, N. I., Torgashovc, G. V., Zakharchenkoc, Yu. F., Fedorovb, E. A.: Val'chukb, V. P.: Electron field emission from nanofilament carbon films. Chem. Phys. Lett. **233**, 63–68 (1995)
13. Chhowalla, M., Ducati, C., Rupesinghe, N.L., Teo, K.B.K., Amaratunga, G.A.J.: Field emission from short and stubby vertically aligned carbon nanotubes. Appl. Phys. Lett. **79**, 2079–2081 (2001)
14. Choi, W.B., Chung, D.S., Kang, J.H., Kim, H.Y., Jin, Y.W., Han, I.T., Lee, Y.H., Jung, J.E., Lee, N.S., Park, G.S., Kim, J.M.: Fully sealed, high-brightness carbon-nanotube field-emission display. Appl. Phys. Lett. **75**, 3129–3131 (1999)
15. Cooper, E.B., Manalis, S.R., Fang, H., Dai, H., Matsumoto, K., Minne, S.C., Hunt, T., Quate, C.F.: Terabit-per-square-inch data storage with the atomic force microscope. Appl. Phys. Lett. **75**, 3566–3568 (1999)
16. Cui, Y., Zou, Y., Valfells, M., Reiser, M., Alter, M., Haber, I., Kishek, R.A., Bernal, S., O'Shea, P.G.: Design and operation of a retarding field energy analyzer with variable focusing for space-charge-dominated electron beams. Rev. Sci. Instrum. **75**, 2736–2745 (2004)
17. Cumings, A., Zettl, A.: Low-friction nanoscale linear bearing realized from multiwall carbon nanotubes. Science **289**, 602–604 (2000)
18. Cumings, J., Zettl, A., McCartney, M.R., Spence, J.C.H.: Electron holography of field-emitting carbon nanotubes. Phys. Rev. Lett. **88**, 056804 (2002)
19. De Heer, W.A., Chatelain, A., Ugarte, D.: A carbon nanotube field-emission electron source. Science **270**, 1179–1180 (1995)
20. De Jonge, N.: The brightness of carbon nanotube electron emitters. J. Appl. Phys. **95**, 673–681 (2004)
21. De Jonge, N., Lamy, Y., Schoots, K. and Oosterkamp, T.H.: High brightness electronbeam from a multi-walled carbon nanotube. Nature. 420, 393–395 (2002)
22. De Jonge, N., van Druten, N.J.: Field emission from individual multiwalled carbon nanotubes prepared in an electron microscope. Ultramicroscopy **95**, 85–91 (2003)
23. De Jonge, N., Bonard, J.M.: Carbon nanotube electron sources and applications. Phil. Trans. R. Soc. Lond A. **362**, 2239–2266 (2004)
24. De Jonge, N., Allioux, M., Doytcheva, M., Kaiser, M., Teo, K.B.K., Lacerda, R.G., Milne, W.I.: Characterization of the field emission properties of individual thin carbon nanotubes. Appl. Phys. Lett. **85**, 1607–1609 (2004)
25. De Jonge, N., Doytcheva, M., Allioux, M., Kaiser, M., Mentick, S.A.M., Teo, K.B.K., Lacerda, R.G., Milne, W.I.: Cap closing of thin carbon nanotubes. Adv. Mater. **17**, 451–455 (2005)
26. De Pablo, P.J., Howell, S., Crittenden, S., Walsh, B., Graugnard, E., Reifenberger, R.: Correlating the location of structural defects with the electrical failure of multiwalled carbon nanotubes. Appl. Phys. Lett. **75**, 3941–3943 (1999)

27. Dean, K.A., Chalamala, B.R.: Field emission microscopy of carbon nanotube caps. J. Appl. Phys. **85**, 3832–3836 (1999a)
28. Dean, K.A., Chalamala, B.R.: The environmental stability of field emission from singlewalled carbon nanotubes. Appl. Phys. Lett. **75**, 3017–3019 (1999b)
29. Dean, K.A., von Allmen, P., Chalamala, B.R.: Three behavioral states observed in field emission from single-walled carbon nanotubes. J. Vac. Sci. Technol., B **17**, 1959–1968 (1999a)
30. Dean, K.A., Groening, O., Kuttel, O.M., Schlapbach, L.: Nanotube electronic states observed with thermal field emission electron spectroscopy. Appl. Phys. Lett. **75**, 2773–2775 (1999b)
31. Dean, K.A., Chalamala, B.R.: Current saturation mechanisms in carbon nanotube field emitters. Appl. Phys. Lett. **76**, 375–377 (2000)
32. Dean, K.A., Burgin, T.P., Chamala, B.R.: Evaporation of carbon nanotubes during electron field emission. Appl. Phys. Lett. **79**, 1873–1875 (2001)
33. Dean, K.A., Chalamala, B.R.: Experimental studies of the cap structures of single-walled carbon nanotubes. J. Vac. Sci. Technol. B **21**, 868–871 (2003)
34. den Engelsen, D., Li, X., Qi, Y.: Properties of a scanning field emission backlight. In: Proceedings of 13th International Display Workshops (IDW'06), Otsu, Japan, 6–8 Dec 2006
35. den Engelsen, D., Silver, J., Withnall, R., Ireland, T. G., Harris, P.G.: Does a Field Emission Backlight make Sense? In: Proceedings of the 16th International Display Workshop (IDW'09), Miyazaki, Japan, 9–11 Dec 2009
36. Dyke, W.P., Dolan, W.W.: Advances in electronics and electron physics. **8**, 89–157 (1956)
37. Fan, S., Chapline, M.G., Franklin, N.R., Tombler, T.W., Cassell, A.M., Dai, H.: Self-oriented regular arrays of carbon nanotubes and their field emission properties. Science, **283**, 512–514 (1999)
38. Franklin, A.D., Luisier, M., Han, S.J., Tulevski, G., Breslin, C.M., Gignac, L., Lundstrom, M.S., Haensch, W.: Sub-10 nm carbon nanotube transistor. Nano. Lett. **12**, 758–762 (2012)
39. Fransen, M.J., van Rooy, T.L., Kruit, P.: Field emission energy distributions from individual multiwalled carbon nanotubes. Appl. Surf. Sci. **146**, 312–327 (1999)
40. Gadzuk, J.W., Plummer, E.W.: Field emission energy distribution (FEED). Rev. Mod. Phys. **45**, 487–548 (1973)
41. Gomer, R.: Field emission and field ionization. Harvard University Press, Cambridge (1961)
42. Groening, O., Kuettel, O.M., Schaller, E., Groening, P., Schlapbach, L.: Vacuum arc discharges preceding high electron field emission from carbon films. Appl. Phys. Lett. **69**, 476–478 (1996)
43. Groening, O., Kuettel, O.M., Emmenegger, C., Groening, P., Schlapbach, L.: Field emission properties of carbon nanotubes. J. Vac. Sci. Technol. B **18**, 665–678 (2000)
44. Hainfeld, J.F.: Understanding and using field emission sources. Scan. Electron Microsc. **1**, 591–604 (1977)
45. Harris, P.J.F.: Carbon Nanotubes and Related Structures: New Materials for the Twenty-First Century. Cambridge University press, New York (1999)
46. Hata, K., Takakura, A., Saito, Y.: Field emission microscopy of adsorption and desorption of residual gas molecules on a carbon nanotube tip. Surf. Sci. **490**, 296–300 (2001)
47. Hawkes, P.W., Kasper, E.: Applied geometrical optics. Principles of electron optics, vol. II (1996)
48. Huang, S., Mau, A.W.H., Turney, T.W., White, P.A., Dai, L.: Patterned growth of wellaligned carbon nanotubes: A soft-lithographic approach. J. Phys. Chem. B. **104**, 2193–2196 (2000)
49. Iijima, S.: Helical microtubulus of graphite carbon. Nature **354**, 56–58 (1991)
50. Jung, I.S., Seonghoon, L., Yoon, H.S., Sung, Y.C., Kyoung, I., Kee, S.N.: Patterned selective growth of carbon nanotubes and large field emission from vertically well-aligned carbon nanotube field emitter arrays. Appl. Phys. Lett. **78**, 901–903 (2001)
51. Kaempgen, M., Chan, C.K., Ma, J., Cui, Y., Gruner, G.: Printable thin film super capacitors using single-walled carbon nanotubes. Nano. Lett. **9**, 1872–1876 (2009)

52. Khazaei, M., Dean, K.A., Farajian, A.A., Kawazoe, Y.: Field emission signature of pentagons at carbon nanotube caps. J. Phys. Chem. **111**, 6690–6693 (2007)
53. King, P.J., Higgins, T.M., De, S., Nicoloso, N., Coleman, J.M.: Percolation Effects in Super capacitors with thin, transparent carbon nanotube electrodes. ACS Nano **6**, 1732–1741 (2012)
54. Kim, C., Kin, B., Lee, S.M., Jo, C., Lee, Y.H.: Eletronic structures of capped carbon nanotubes under electric fields. Phys. Rev. B. **65**, 165418-1-165418-6 (2002)
55. Küttel, O.M., Gröning, O., Emmenegger, C., Schlapbach, L.: Electron field emission from phase pure nanotube films grown in a methane/hydrogen plasma. Appl. Phys. Lett. **73**, 2113–2115 (1998)
56. Kuzumaki, T., Takamure, Y., Ichinose, H., Horiike, Y.: Structural change at the carbonnanotube tip by field emission. Appl. Phys. Lett. **78**, 3699–3701 (2001)
57. Leopold, J.G., Zik, O., Cheifetz, E., Rosenblatt, D.: Carbon nanotube-based electron gun for electron microscopy. J. Vac. Sci. Technol. A **19**, 1790–1795 (2001)
58. Lovall, D., Buss, M., Graugnard, E., Andres, R.P., Reifenberger, R.: Electron emission and structural characterization of rope of single-walles carbon nanotubes. Phys. Rev. B. **61**, 5683–5691 (2000)
59. Ma, X., Wang, E., Wuzong, Z., Jefferson, D.A., Jun, C., Shaozhi, D., Ningsheng, X., Jun, Y.: Polymerized carbon nanobells and their field-emission properties. Appl. Phys. Lett. **75**, 3105–3107 (1999)
60. Mann, M., El Gomati, M., Wells, T., Milne, W.I., Teo, K.B.K.: The application of carbon nanotube electron sources to the electron microscope. Proc. Of Spie. **7037**, 7037P-1-7037P-6 (2008)
61. Nilsson, L., Groening, O., Emmenegger, C., Kuettel, O., Schaller, E., Schlapbach, L., Kind, H., Bonard, J.M., Kern, K.: Scanning field emission from patterned carbon nanotube films. Appl. Phys. Lett. **76**, 2071–2073 (2000)
62. Nishikawa, O., Tomitori, M., Iwawaki, F.: High resolution tunneling microscopies: from FEM to STS. Surf. Sci. **266**, 204–213 (1992)
63. Obraztsov, A.N., Volkov, A.P., Pavlovskii, I.Y., Chuvilin, A.L., Rudina, N.A., Kuznetsov, V.L.: Role of the curvature of atomic layers in electron field emission from graphitic nanostructured carbon. JETP Lett. **69**, 411–417 (1999)
64. Obraztsova, E.D., Bonard, J.M., Kuznetsov, V.L., Zaikovskii, V.I., Pimenov, S.M., Pozarov, A.S., Terekhov, S.V., Konov, V.I., Obraztsov, A.N., Volkov, A.S.: Structural measurements for single-wall carbon nanotubes by Raman scattering technique. Nanostruct. Mater. **12**, 567–572 (1999)
65. Purcell, S.T., Vincent, P., Journet, C., Binh, V.T.: Hot nanotubes: stable heating of individual multiwall carbon nanotubes to 2000 K induced by the field-emission current. Phys. Rev. Lett. **88**, 105502 (2002)
66. Rinzler, A.G., Hafner, J.H., Nikolaev, P., Lou, L., Kim, S.G., Tomanek, D., Nordlander, P., Colbert, D.T., Smalley, R.E.: Unraveling nanotubes: field emission from an atomic wire. Science **269**, 1550–1553 (1995)
67. Rosen, R., Simendinger, W., Debbault, C., Shimoda, H., Fleming, L., Stoner, B., Zhou, O.: Application of carbon nanotubes as electrodes in gas discharge tubes. Appl. Phy. Lett. **76**, 1668–1670 (2000)
68. Ryhänen, T., Uusitalo, A.M., Ikhala, O., Kärkkäinen, A.: New Technology for Flat Panel Displays. In Nanotechnologies for Future Mobile Devices. Cambridge University press, New York (2010). 221
69. Saito, R., Fujita, M., Dresselhaus G., Dresselhaus, M. S.: Electronic structure of chiral graphene tubules. Appl. Phys. Lett. **60**, 2204–2206 (1992)
70. Saito, R., Dresselhaus, G., Dresselhaus, M.S.: Physical Properties of Carbon Nanotubes. Imperial College Press, London (1998)
71. Saito, Y., Uemura, S.: Field emission from carbon nanotubes and its application to electron sources. Carbon **38**, 169–182 (2000)

72. Saito, Y., Hata, K., Murata, T.: Field emission patterns originating from pentagons at the tip of a carbon nanotube. Jpn. J. Appl. Phys. **39**, L271–L272 (2000)
73. Semet, V., Binh, V.T., Vincent, P., Guillot, D., Teo, K.B.K., Chhowalla, M., Amaratunga, G.A.J., Milne, W.I., Legagneux, P., Pribat, D.: Field electron emission from individual carbon nanotubes of a vertically aligned array. Appl. Phys. Lett. **81**, 343–345 (2002)
74. Sinha, N., Ma, J., Yeow, J.T.W.: Carbon Nanotube-based Sensors. J. Nanosci. Nanotechnol. **6**, 573–590 (2006)
75. Spindt, C.A.: A thin film field emission cathode. J. Appl. Phys. **39**, 3504–3505 (1968)
76. Sugie, H., Tanemure, M., Filip, V., Iwata, K., Takahashi, K., Okuyama, F.: Carbon nanotubes as electron source in an X-ray tube. Appl. Phys. Lett. **78**, 2578–2580 (2001)
77. Swanson, L.W., Schwind, G.A.: A review of the ZrO/W Schottky cathode". In Handbookof charged particle optics (ed. J. Orloff). 77–102. CRC Press, Boca Raton, (1997)
78. Takakura, A., Hata, K., Saito, Y., Matsuda, K., Kona, T., Oshima, C.: Energy distributions of field emitted electrons from a multi-wall carbon nanotube. Ultramicroscopy **95**, 139–143 (2003)
79. Teo, K.: Carbon nanotube electron source technology. JOM **59**, 29–32 (2007)
80. Van Veen, A.H.V., Hagen, C.W., Barth, J.E., Kruit, P.: Reduced brightness of the Zr/W Schottky electron emitter. J. Vac. Sci. Technol. B **19**, 2038–2044 (2001)
81. Wang, Z.L., Poncharal, P., de Heer, W.A.: In situ imaging of field emission from individual carbon nanotubes and their structural damage. Appl. Phys. Lett. **80**, 856–858 (2002)
82. Wei, Y.Y., Dean, K.A., Coll, B.F., Jaskie, J.E.: Stability of carbon nanotubes under electric field studied by scanning electron microscopy. Appl. Phys. Lett. **79**, 4527–4529 (2001)
83. Wildoer, J.W.G., Venema, L.C., Rinzler, A.G., Smalley, R.E., Dekker, C.: Electronic structure of atomically resolved carbon nanotubes. Nature **391**, 59–62 (1998)
84. Wong, S.S., Joselevich, E., Woolley, A.T., Li, C.C., Lieber, C.M.: Covalently functionalized nanotubes as nanometresized probes in chemistry and biology. Nature **394**, 52–55 (1998)
85. Wu, Z., Chen, Z., Du, X., Logan, J.M., Sippel, J., Nikolou, M., Kamaras, K., Reynolds, J.R., Tanner, B.D., Hebard, A.F., Rinzler, A.G.: Transparent, conductive carbon nanotube films. Sci. **305**, 1273–1276 (2004)
86. Yaguchi, T., Sato, T., Kamino, T., Taniguchi, Y., Motomiya, K., Tohji, K., Kasuya, A.: A method for characterizing carbon nanotubes. J. Electron Microsc. **50**, 321–324 (2001)
87. Yue, G.Z., Qiu, Q., Gao, B., Cheng, Y., Zhang, J., Shimoda, H., Chang, S., Lu, J.P., Zhou, O.: Generation of continuous and pulsed diagnostic imaging X-ray radiation using a carbon-nanotube-based field-emission cathode. Appl. Phys. Lett. **81**, 355–357 (2002)
88. Yamamoto, S., Watanabe, I., Sasaki, S., Yaguchi, T.: Absolute work function measurements with the retarding potential method utilizing a field emission electron source. Surf. Sci. **266**, 100–106 (1992)
89. Yu, M. F., Lourie, O., Dyer M.J., Moloni, K., Kelly, T.F., Ruo, R.S.: Strength and breaking mechanism of multiwalled carbon nanotubes under tensile load. Science. 287, 637–640 (2000)
90. Xu, D., Guo, G., Gui, L., Tang, Y., Shi, Z., Jin, Z., Gu, Z., Liu, W., Li, X., Zhang, G.: Controlling growth and field emission property of aligned carbon nanotubes on porous silicon substrates. Appl. Phys. Lett. **75**, 481–483 (1999)
91. Xu, N.S., Huq, S.E.: Novel cold cathode materials and applications. Mater. Sci. Eng. R **48**, 47–189 (2005)
92. Zhang, D., Ryu, K., Liu, X., Polikarpov, E., Ly, J., Tompson, M.E., Zhou, C.: Transparent, conductive, and flexible Carbon nanotube films and their application in Organic light-emitting diodes. Nano Lett. **6**, 1880–1886 (2006)
93. Zhang, T., Mubeen, S., Myung, N.V., Deshusses, M.A.: Recent progress in carbon nanotube-based gas sensors. Nanotechnology **19**, 1–14 (2008)
94. Zhirnov, V.V., Lizzu- Rinne, C., Wojak, G.J., Sanwald, R.C., Hren, J.J.: Standardization of field emission measurements. J. Vac. Technol. B **19**, 87–93 (2001)

95. Zhao, Y., Wei, J., Vajtai, R., Ajayan, P.M., Barrera, E.V.: Iodine doped carbon nanotube cables exceeding specific electrical conductivity of metals".Sci. Rep. **83**, 1–5 (2011)
96. Zhou, O., Fleming, R.M., Murphy, D.W., Chen, C.H., Haddon, R.C., Ramirez, A.P., Glarum, S.H.: Defects in carbon nanostructures. Science **263**, 1744 (1994)
97. Zhu, W., Bower, C., Zhou, O., Kochanski, G., Jin, S.: Large current density from carbon nanotube field emitters. Appl. Phys. Lett. **1999**(75), 873–875 (1999)
98. Zhu, W.: Vacuum Micro-Electronics. Wiley, New York (2001)

Synthesis and Characterisation of Carbon Nanocomposites

M. Z. Krolow, C. A. Hartwig, G. C. Link, C. W. Raubach,
J. S. F. Pereira, R. S. Picoloto, M. R. F. Gonçalves,
N. L. V. Carreño and M. F. Mesko

Abstract Carbon nanocomposites have received more attention in the last years in view of their special properties such as low density, high specific surface area, and thermal and mechanical stability. Taking into account the importance of these materials, many studies aimed at improving the synthesis process have been conducted. However, the presence of impurities could affect significantly the properties of these materials, and the characterisation of these compounds is an important challenge to assure the quality of the new carbon nanocomposites. Thus, in this work are presented the characteristics of carbon nanocomposites, the improvements and developments in the synthesis process, as well as the most used characterisation techniques of these compounds.

M. Z. Krolow · G. C. Link · M. R. F. Gonçalves · N. L. V. Carreño
Centro de Desenvolvimento Tecnológico, Universidade Federal de Pelotas, Pelotas-RS,
96010-900, Brazil

C. A. Hartwig · M. F. Mesko (✉)
Centro de Ciências Químicas, Farmacêuticas e de Alimentos, Universidade Federal de
Pelotas, Pelotas-RS, 96010-900, Brazil
e-mail: marciamesko@yahoo.com.br

C. W. Raubach
LIEC, Departamento de Química, Universidade Federal de São Carlos, São Carlos-SP,
13565-905, Brazil

J. S. F. Pereira · R. S. Picoloto
Departamento de Química, Universidade Federal de Santa Maria, Santa Maria-RS, Brazil

C. Avellaneda (ed.), *NanoCarbon 2011*, Carbon Nanostructures,
DOI: 10.1007/978-3-642-31960-0_2, © Springer-Verlag Berlin Heidelberg 2013

1 General Aspects of Carbon Nanocompounds

Nanomaterials (NMs) are defined as materials that have structural features with at least one dimension of 100 nm or less; they include nanofilms and nanocoatings (one dimension), nanotubes and nanowires two dimensions) and nanoparticles (three dimensions) [1]. Nanomaterials may present itself in different sizes and shapes, and also differ in composition and origin. Depending on the interaction between the nanoparticles, the NMs can be found as single particles, aggregates, powders or dispersed in a matrix, over colloids, suspensions and emulsions, nanolayers and films, and coated or stabilised as fullerenes and their derivates [2].

Carbon-based nanomaterials, including fullerenes, single- and multi-walled carbon nanotubes and carbon nanoparticles, are currently one of the most attractive nanomaterials from an applications perspective [3]. Since their discovery in 1991 by Sumio Iijima, carbon nanotubes have been intensively studied [4]. Their extraordinary electronic and mechanic properties point towards a great variety of potential future applications, including polymer composites [5], electronics [6] and drug delivery [3, 7–9].

Carbon nanotubes (CNTs), one of most researched carbon-based materials, have a unique atomic structure, very high aspect ratio and extraordinary mechanical properties (strength and flexibility), making them ideal reinforcing fibers in nanocomposites [10]. The effective utilisation of carbon nanotubes in composite applications depends strongly on the ability to disperse the nanotubes homogeneously throughout the matrix, without destroying the integrity of the nanotubes. To take advantage of the superior mechanical properties of the carbon nanotubes, they have been used as fillers in epoxy resins [11] and in the production of polymer nanocomposites [12, 13]. On the other hand, the cost of production of carbon nanofibers (CNFs) is significantly less than CNTs, and could be advantageous when compared to CNTs in some applications. The dimensions of CNFs are between 50 and 200 nm [14], which are similar to those of single-walled carbon nanotubes (SWCNTs) and multi-walled carbon nanotubes (MWCNTs) [15]. The main characteristic that distinguishes CNFs from CNTs resides in graphene plane alignment. If the graphene plane and fiber axis do not align, the structure is defined as a CNF, but when both are parallel, the structure is considered a CNT [16].

Other carbon compounds extensively studied are the carbon-containing nanoparticles, usually called carbon nanocomposites [17–21]. These materials have received great attention due to properties such as low density, high specific surface area and uniform pore size, and thermal and mechanical stability [22]. It is important to note that the differential application of this type of material is in the properties of the particles that are associated, such as magnetic, catalytic, energy and absorptive, among others [22, 23]. An image of typical nanocomposite metal/carbon obtained by means of transmission electron microscopy (TEM) can be observed in Fig. 1 [24].

Nanocomposites consisting of metal and carbon are generally produced using techniques that involve heat treatment, during which some of the carbon may or

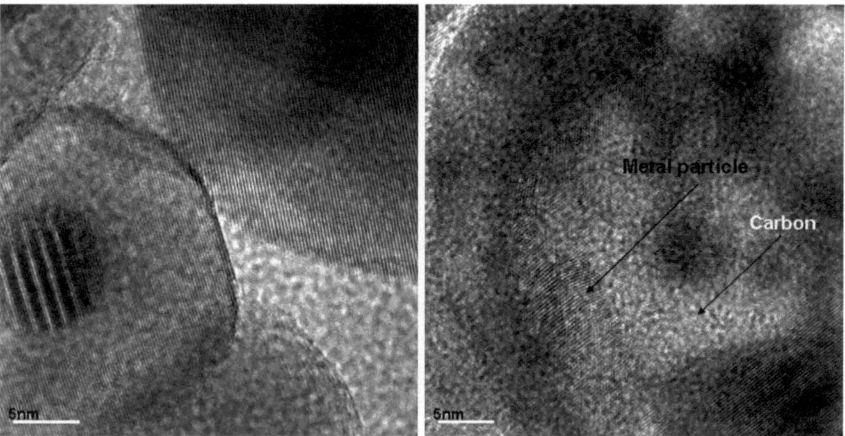

Fig. 1 TEM images of Ni/C catalyst before catalytic test. Copyright (2011), with permission from Elsevier

may not be eliminated. It is also at this stage that the metal nanoparticles are formed, and the material's crystallinity is increased. Nanoparticles often have spherical shapes with diameters between 10 and 20 nm as well as high crystallinity. These nanoparticles are very stable because they are embedded in the carbon matrix. Depending on the type of treatment they have undergone, the nanoparticles surface area can be quite variable, with values ranging from 50 m^2/g to 1000 m^2/g. Also, depending on the method of synthesis used or application for the product, the metal content can also vary greatly [24–29].

2 Synthesis of Carbon Nanocomposites

The synthesis of nanomaterials has been widely studied, mainly in regard to carbon nanotubes in view of the desire to obtain CNTs with high quality and at low cost. However, in recent years, many carbon compounds containing nanoparticles also have been studied and, as mentioned before, in many cases these compounds could present some advantages in several industrial applications. In this way, many studies on the synthesis process have been performed with the goal of obtaining compounds with good characteristics using techniques that take less time and require less-toxic reagents.

Carbon-based nanocomposites containing nanoparticles can be synthesised in several ways. The supports for the carbon nanoparticles can be produced, for example, using silica matrices [30–32], also through deposition, the CVD method, on the carbon mold [22]. Then, to remove the silica, the material is subjected to a solution containing hydrofluoric acid. In such cases, which produce a carbon matrix in the first step, the nanoparticles are usually introduced into the pores of the matrix by wet impregnation and the composites are subjected to heat treatment. On the other hand, the carbon matrix and the nanoparticles can be produced in situ by, for example, the polymeric precursor method, used for both types of matrices, silica-based and carbon-based [24, 26, 33].

A system of nanomaterials that has received great attention lately is the so-called core–shell system, which consists of nanoparticles that form a core, the surface of which is coated with a thin layer of another material. Usually, this system has a core of a metal or oxide and a shell made of a carbonaceous or polymeric material, or even other oxides or metals. In these systems, the shell provides a layer that protects the core against oxidation, as in the case of nano-crystals of Fe (core)/iron oxide (shell) or Fe (core)/Au (shell), or to provide reactive capacity that does not represent the core alone [27, 34].

There are several types of synthetic routes for the synthesis of nanocomposites, among which are impregnations, the polymeric precursor, CVD, ball milling and sol–gel synthesis [24, 27, 35–37]. More than one method can be combined in different stages of the formation of nanocomposites, such as solvothermal and hydrothermal co-precipitation, or they can be combined with the reduction of metal through the use of a reducing agent [38, 39].

The method of wet impregnation can be used when one of the components of the composite, such as a silica matrix, for example, is already available. Normally, this procedure is performed in a solvent in which the separation of the particles is easy, such as a low molecular weight alcohol. Seo and co-workers [27] impreg-nated cations of Fe and Co into silica powder of high surface area. Later, they were reduced to a metallic state, at high temperature and atmosphere of H_2. The pro-cedure was followed by deposition of a thin carbon layer, from methane decom-position on the FeCo nanoalloy, by CVD method, methane decomposed on the FeCo nanoalloy. To make the compounds water-soluble, a non-covalent func-tionalisation step was applied using phospholipid-poly (ethylene glycol). The materials showed excellent magnetic activity and high stability, with applicability as a contrast agent [27].

The polymeric precursor method is widely used when to form a matrix containing well-dispersed nanoparticles, based on the Pechini [40] procedure. This method, involves the complexation of metal cations with a complexing agent such as citric acid and subsequent polymerisation by polyesterification reactions, using, for example, ethylene glycol as the polymerising agent. Carreño and colleagues devel-oped a catalyst metal/carbon via the polymeric precursor route. After the synthesis of polymer resin, the material was calcined at high temperatures under inert N_2 atmo-sphere. In breach of the organic chains, the reducing atmosphere of CO formed was able to reduce the metals to the metallic state, thus forming composite metal nano-particles in an amorphous carbon matrix in which nanoparticles of easily oxidised metals such as nickel and cobalt remained protected from oxidation [24, 41].

High-energy milling is a generic term that can be used to describe various grinding processes, such as mechanical alloying, mechanical milling or mechano-chemical process, depending on the precursor powders used in the mix. In all cases, the activation process is mechanical. In this case, not only the mechanical friction is important, but also the main reaction mechanism that is related to collision between the balls and powder processed. The grinding process is usually performed using a solvent such as isopropyl alcohol. This method is widely applied in the development of nanostructured metals and alloys and of ceramics [40, 42–44].

The sol–gel method of synthesis is very attractive for the synthesis of nanostructures containing more than one component, since the slow reaction kinetics allow good structural engineering of the final product. Another advantage is that the reactions are conducted at low temperatures or at room temperature. The sol–gel process involves inorganic precursors that undergo various chemical reactions, resulting in the formation of a three-dimensional molecular network. One of the most common routes is via hydrolysis and condensation of metal alkoxides to form larger metal oxide molecules that polymerize to form the coating. The sol–gel procedure allows coating of substrates with complex shapes on the nanometer to micrometer scale, which some commonly used coating procedures cannot achieve. The substrates include colloidal particles, organic/inorganic crystals, or even fibers and nanotubes [45–47].

The sol–gel method is also widely used for the synthesis of silica matrices. The silicon alkoxides are the most commonly used to create a sol–gel matrix such as the SiO_2/C prepared by Gushikem and colleagues [36], or even to enclose a network of nanoparticles in the xerogel [48].

Nanocomposites containing carbon nanotubes have been explored later. The addition of these nanotubes to various matrices, such as metal, for example, can improve the mechanical properties, like stiffness, wear and fatigue, and electrical properties in comparison to matrices without nanotubes [49]. Some years ago, it was observed the low wetting of the nanotubes in a metal matrix in the case of composites synthesised by hot pressing [50, 51]. Alternatively, powder coating processes can be employed with great success, making the distribution of carbon nanotubes more homogeneous. The chemical coating of carbon nanotubes by loads of metal nanoparticles has been widely used [52]. In this whole context, it is easy to see the importance of analytical control of these loads present in the nanotubes, as well as the proportions of metal/nanotube nanocomposites for the formation of technologically relevant materials, since small variations in these parameters can make big changes in the properties of the final product.

Graphene oxide (GO) nanosheets impregnated with silver nanoparticles (Ag NPs) were fabricated by the in situ reduction of adsorbed Ag + by hydroquinone in a citrate buffer solution. Paper-like Ag NP/GO composite materials were fabricated owing to convenient structure characterisation and antibacterial tests. Antibacterial activity was tested using *Escherichia coli* and *Staphylococcus aureus* as model strains of Gram negative and Gram positive bacteria, respectively. The as-prepared composites exhibit stronger antibacterial activity against both. The Ag NP/GO composites performed efficiently in bringing down the count of *E. coli* from 106 cfu/mL to zero with 45 mg/L GO in water. The micron-scale GO nanosheets (lateral size) enable them to be easily deposited on porous ceramic membranes during water filtration, making them a promising biocidal material for water disinfection. GO nanosheets can be viewed in the Fig. 2 [53].

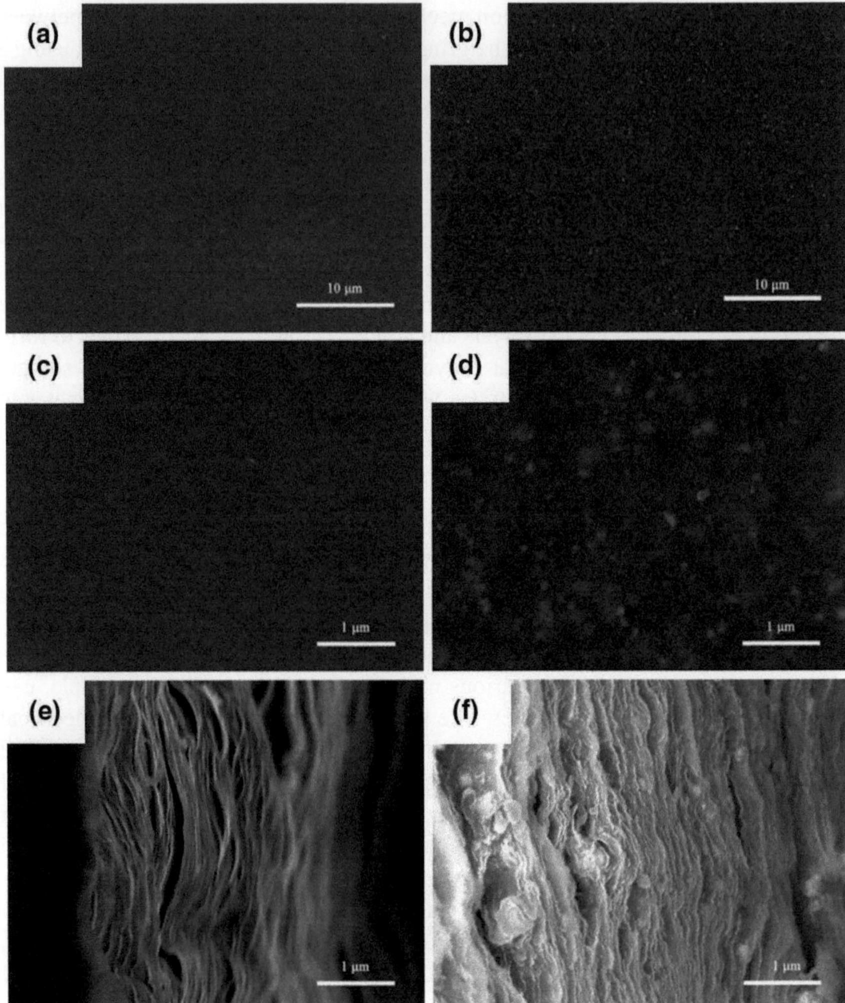

Fig. 2 *Top* view FESEM images of (**a** and **c**) GO paper and (**b** and **d**) Ag NP/GO composite paper. Side-view FESEM images of the cross sections of (**e**) GO paper and (**f**) Ag NP/GO composite paper. Copyright (2011), with permission from Elsevier

In other work, the authors synthesised one-dimensional rod-like nickel nanostructure fabricated through a simple, efficient and one-pot solvothermal approach with hydrazine hydrate and trimethylamine as reducing and morphology-directing agents. Magnetisation studies showed that the nanorod presents a distinct enhanced coercive force as a reflection of the 1 D nanostructure. Because of the high coercive force, we believe that the as-obtained Ni nanorod is suitable material for potential applications in magnetic storage devices and catalysis [54].

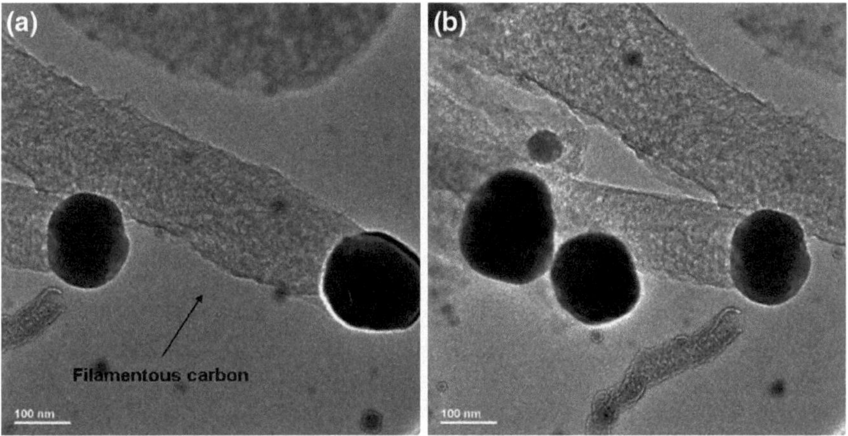

Fig. 3 TEM images of Ni/C catalyst after catalytic test at 500 °C. Copyright (2009), with permission from Elsevier

3 Applications of Carbon Nanocomposites

3.1 Catalytic Applications

Materials of various matrices containing various transition metals have been widely applied as catalysts [55, 56]. In the field of catalysts based on carbon, Tavasoli and coauthors [37] reported the synthesis of Co/CNT by sequential impregnation method, in order to investigate the interesting catalytic properties of this composite display. They varied the cobalt loadings from 15 to 40 wt %. The authors assessed the physicochemical characteristics and catalytic performance for Fischer–Tropsch synthesis and compared the results with an alumina-supported cobalt catalyst. The results showed that the hydrocarbon yield obtained by the CNT-supported cobalt catalyst is surprisingly much larger than those obtained from cobalt on other inorganic supports, and moreover the CNT caused a slight decrease in the reaction product distribution to lower molecular weight hydrocarbons.

Carreño et al. [24] synthesised nickel-carbon nanocomposites for use as catalysts for ethanol steam reforming. The authors proposed the use of castor oil as a carbon precursor in the synthesis. The results showed that nickel/carbon catalysts have a high activity for ethanol steam reforming. The authors concluded that catalytic activity was increased at high reaction temperatures, which may be associated with the formation of filamentous carbon. Images of nickel/carbon catalyst after their performance are displayed in Fig. 3.

It is important to note that this work shows an economical and environmentally friendly alternative to preparing catalysts from renewable resources, like castor oil, that are a low-cost, biomass-derived feedstock. This oil is obtained from extracting of the plant Ricinus communis, of the family *Euphorbiaceae* that grows in large quantities in most tropical and sub-tropical countries, such as Brazil [57].

3.2 Sensors Applications

Because of their excellent electrochemical properties, such as rapid electron kinetics, semi- and superconducting electron transport, high tensile strength composites, and hollow core suitable for storing guest molecules, CNTs have attracted attention as an electrode material for electrochemical sensors. Associating CNTs and conducting polymer with synergistic effect it's possible improve the electrical and mechanical properties of polymers in order to develop high performance sensor [4, 58].

Very recently, a sensor for determination of creatinine was produced. The biosensor was produced from a mixture of creatinine amidohydrolase, creatine amidino-hydrolase and sarcosine oxidase coimmobilised via N-ethyl-N0-(3-dimethylaminopropyl) carbodiimide and N-hydroxy succinimide onto nanocomposite films of carboxylated-NTC/polyaniline electrodeposited on the surface of a platinum electrode. The sensor could detect creatinine in levels as low as 0.1 μM. The results showed that the use of a NTC/polyaniline composite to construct a creatinine biosensor led to improved analytical performance of the sensor, which required low power and possessed fast response times, high sensitivity and storage stability. The authors also suggested, based on the results, that this composite could be used to improve the performance of various other types of biosensors [59].

3.3 Magnetic Applications

Li et al. [38] studied uniform nitrogen-enriched, carbon-encapsulated nickel nanospheres obtained by a novel solvothermal method. The spheres as-prepared presented a core–shell structure with a nickel core surrounded by 10 nm thickness nitrogen-enriched carbon shell. Samples showed ferromagnetic behavior, due to the metallic nickel core. The introduction of the nitrogen element resulted in the surface modification of the encapsulating metal nanospheres, and this material, according to the authors, may play an important role in biomedical or other applications.

Sunny et al. [60] developed a simple and low-cost technique for obtaining highly stable, carbon-coated, nickel nanostructures at relatively low reaction temperatures. The nickel nanoparticles were obtained by a sodium borohydride reduction technique and coated with oleic acid. The samples were subjected to calcination at 873 K, in high purity nitrogen flow. This method obtained face-centered, cubic nickel nanoparticles with 5 nm thick carbon layers. The large saturation magnetisation combined with high stability make the nanoparticles suitable for use in functionalised drug targeting, as solid lubricants and electromagnetic shield materials (Fig. 4).

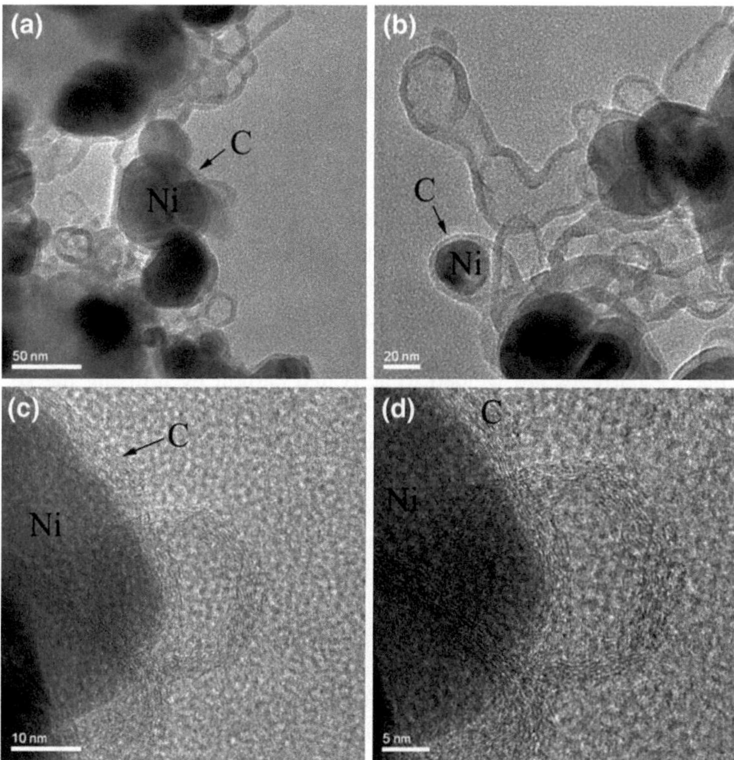

Fig. 4 TEM image (**a–b**) and HRTEM image (**c–d**) of Ni@C. Copyright (2009), with permission from Elsevier

Sunny et al. [39] still studied magnetic nickel/carbon hybrid nanostructures like a novel material with microwave absorbing properties. It was found that hybrid nanostructures are an effective microwave absorber with reflection loss less than − 10 dB in the S and X-bands. The space charge polarization from the nickel carbon interface and intrinsic magnetic loss of the hybrid magnetic metal carbon nanostructures cause enhanced microwave absorbing properties

Powdered activated carbon (AC) has been the standard adsorbent for the reclamation of residential and industrial wastewaters for the recent decades [61, 62]. Associating these properties with magnetic properties, magnetic activated carbon (MAC) has been used to remove organic and inorganic pollutants [63, 64]. One recent example of these studies is the comparison between AC and MAC with regard to imidacloprid adsorption. MAC powder was formed from an iron oxide/ AC composite. In conclusion, the authors observed less surface area and micropore volume for the magnetic powder; however, it was concluded that MAC powder can be successfully applied for the removal of environmental pollutants [65].

3.4 Biomedical Applications

Magnetic drug delivery, as proposed by Widder et al. [66], is a method by which a magnetically susceptible material coated with a drug-laden matrix is injected and then an externally placed magnet is used to guide the drug matrix to the target site. Several types of ferromagnetic materials have been used as matrices for application in drug delivery systems, as well as the magnetic separation of biomolecules. Composites of ferromagnetic material and activated carbon can be used for such purpose. The ferromagnetic material such as iron provides the magnetic properties, while activated carbon works as a drug-adsorption matrix. One of the methods [35] to prepare the magnetic target carrier is by a high-impact ball milling process in which fine Fe particles are welded to activated carbon by utilising the so-called mechanochemical effect. However, the major disadvantage is that it is easily separated into Fe and activated carbon while being transferred in veins. The first particles used for drug delivery were based on colloidal magnetite coated with cross linked albumin. These microspheres were used by Widder and coworkers [67] to encapsulate doxorubicin, and were captured magnetically in Yoshida sarcoma tumors implanted in rat tails. Since these first publications, several studies on biomedical applications were realised.

Most recently, in the same field of ferromagnetic metal/carbon composites, Rudge and coworkers studied a method to prepare composites of carbon/iron by a milling technique originally developed for making composite powders [68]. The authors obtained particles with size predominantly between 0.5 and 5.0 μm, which absorbed and desorbed doxorubicin, a potent chemotherapeutic with a narrow therapeutic index. Chemical analysis showed the particles to be composed primarily of carbon, iron and oxygen, with traces of phosphorous, hydrogen, nitrogen and sulfur, and parts per million of trace metals. This study demonstrates the importance of analytical investigation of systems used in biomedical applications.

Five years later, a carbon composite rich in fcc-Co and enclosed by carbon matrix was used as potential magnetic carrier [28]. The particles, which were made of cobalt and carbon, had high surface area for drug adsorption and sufficient magnetic susceptibility for targeting the composite through the use of an external magnet. Before using the cobalt, the authors attempted to use iron to synthesize the material for ion exchange, but the Fe^{2+} oxidized to Fe^{3+} during the ion exchange. Therefore, cobalt which also shows strong magnetic properties was selected. The Co–C composite formed was a spherical 250-300 μm size particle. The spherical particles were ground to a fine powder.

In the biomedical field, the CNTs also have received great attention. Many applications for CNTs have been proposed, including biosensors, drug and vaccine delivery vehicles and novel biomaterials [69]. CNTs can be used as nanofillers in existing polymeric materials to both dramatically improve mechanical properties and create highly anisotropic nanocomposites [70, 71]. They can also be used to create electrically conductive polymers and tissue engineering constructs with the capacity to provide controlled electrical stimulation [72, 73].

An example of the utility of CNTs in biomedicine is their relatively large length-to-diameter aspect ratio (which can exceed 106, with an average length of 1 mm and diameter of $\sim$ 1 nm) with a very large surface area, which makes CNTs attractive for high sensitivity molecular detection and recognition. Consequently, a large fraction of the CNT surface can be modified with functional groups of various complexities, which would modulate CNTs' in vivo and in vitro behavior [74].

As CNTs are intrinsically not water soluble, modification through chemical functionalisation using suitable dispersants and surfactants can enhance solubility to the range of g/mL4 and is essential for their controlled dispersion. For example, constituent polar molecules can render CNTs soluble, whereas nonpolar moieties make CNTs immiscible. Such processes have proved especially important in that nonsolubilised CNTs have been found to cause cell death in culture [75–77].

Each CNT could be intrinsically different due to limitations on the fabrication of structurally identical CNTs with minimal impurities [55]. Subtle variations in local and overall charge, catalyst residue (typically Fe, Co, and Ni), and length of individual nanotubes are three representative issues that preclude precise use of CNTs in the biomedical sciences.

4 Characterisation of Carbon Nanocomposites

With the advent of nanotechnology for developing new materials with a wide variety of special properties, there is a need for greater control of the characteristics of these compounds and also the impurities that could be present, as these new materials have been used for very different applications in almost all fields of technology. The principal techniques for analysis of the composition and structure of nanocomposites are scanning electron microscopy (SEM), x-ray photoelectron spectroscopy (XPS) and energy dispersive x-ray (EDX) spectroscopy [38]. The EDX technique has been widely used to identify metals and/or contaminants in nanocomposites. Sunny et al. used this technique to verify the presence or absence of contaminants in carbon nanotubes. SEM is used to investigate the morphology of nanocomposites. Using the SEM technique, it is possible to observe the surface defects of the material [38, 39, 60–63].

Another technique that is widely used for characterisation of nanocomposites is XPS [64, 65]. This technique can be applied to detect the presence of the elements in the material surface. Maldonado et al. used the XPS technique to evaluate the compositional and structural properties of carbon nanotubes doped with N_2 [66]. XPS spectroscopy was used to investigate fluorine's interaction with the surfaces of nanocompounds. In this study, a MWCNT, SWCNT and a fiber were fluorinated in F_2- and ClF_3 atmosphere at room temperature [35].

Taking into account that metal impurities can affect the physical, chemical and surface properties of carbon nanocomposites and could complicate or prevent their use for industrial applications, it is important to control this type of contamination [78–81]. In addition, it is important to emphasise that there are many applications of carbon nanocomposites in medicine and related areas, and impurities in these materials could cause many risks to human health [82, 83]. Since the presence of metallic impurities can increase the potential risk of using carbon nanocomposites, it is necessary to develop methods to identify metallic elements in low concentrations in these materials [78, 84].

The techniques employed for the determination of metal impurities include atomic spectrometry, for which the samples are most commonly prepared as aqueous solutions. This poses a problem for the determination of impurities in carbon nanocomposites, since these materials are difficult to dissolve due to their structure [85, 86].

Some authors describe different sample preparation methods for carbon nanotubes, including dry ashing, microwave-induced combustion (MIC) or microwave-assisted wet digestion using a mixture of nitric and perchloric acid or with hydrogen peroxide for subsequent determination of Al, Cu, Cr, Fe, Mn, Mo, Ni, Zn and halogens by inductively coupled plasma optical emission spectrometry (ICP OES), by inductively coupled plasma mass spectroscopy (ICP-MS) and by ion chromatography (IC) [87–89]. Additionally, the authors purposed the use of direct solid sampling electrothermal atomic absorption spectroscopy (DSS-ET AAS) for the determination of Al, Cd, Co, Cr, Cu, Mg, Mn and Pb in SWCNTs and MWCNTs [87]. However, for the determination of metals in carbon nanocomposites, there are no applications in literature, so methods should be developed to improve the quality control of these kinds of materials.

5 Conclusion

The increase in the applications of carbon compounds, especially in industry and medicine, requires the development of suitable synthesis processes and also analytical methods for quality control of these compounds. The techniques used to determine the presence of metals in carbon compounds and the sample preparation technique for quantification of such metals should be selected based on factors such as the time required for analysis, simplicity of the procedure, reagent consumption, sample mass, waste generation and low range limits of detection. Regardless of the chosen methodology, studies in the literature point to the need for the determination of metal contaminants in carbon compounds, since some elements that have been found in relatively high levels can interfere with human health or affect the properties of materials intended for industrial applications.

Acknowledgments The authors are thankful for the financial support of Brazilian research financing institutions: CAPES, CNPq and FAPESP

References

1. Tiede, K., Boxall, A.B.A., Tear, S.P., Lewis, J., David, H., Hassellov, M.: Food additives and contaminants part a-chemistry analysis control exposure and risk. Assessment 25(7), 795–821 (2008)
2. Jiang, J.K., Oberdorster, G., Biswas, P.: J. Nanopart. Res. 11(1), 77–89 (2009)
3. Huczko, A.: Appl. Phys. Mater. Sci. Process. 74(5), 617–638 (2002)
4. Iijima, S.: Nature 354(6348), 56–58 (1991)
5. Miyagawa, H., Misra, M., Mohanty, A.K.: J. Nanosci. Nanotechnol. 5(10), 1593–1615 (2005)
6. Jeong, W., Kessler, M.R.: Chem. Mater. 20(22), 7060–7068 (2008)
7. Bianco, A., Kostarelos, K., Prato, M.: Curr. Opin. Chem. Biol. 9(6), 674–679 (2005)
8. Maynard, A.D., Aitken, R.J., Butz, T., Colvin, V., Donaldson, K., Oberdorster, G., Philbert, M.A., Ryan, J., Seaton, A., Stone, V., Tinkle, S.S., Tran, L., Walker, N.J., Warheit, D.B.: Nature 444(7117), 267–269 (2006)
9. Oberdorster, G., Oberdorster, E., Oberdorster, J.: Environ. Health Perspect. 113(7), 823–839 (2005)
10. Wang, Z.l., Poncheral, P., de Heer, W.A.: in Nanostrucutres Systems (1999), 14–18
11. Ajayan, P.M., Stephan, O., Colliex, C., Trauth, D.: Science 265(5176), 1212–1214 (1994)
12. Bower, C., Rosen, R., Jin, L., Han, J., Zhou, O.: Appl. Phys. Lett. 74(22), 3317–3319 (1999)
13. Jin, L., Bower, C., Zhou, O.: Appl. Phys. Lett. 73(9), 1197–1199 (1998)
14. Ku, B.K., Emery, M.S., Maynard, A.D., Stolzenburg, M.R., McMurry, P.H.: Nanotechnology 17(14), 3613–3621 (2006)
15. Wang, H., Xu, Z., Eres, G.: Appl. Phys. Lett. 88(21), 213111 (2006)
16. International Standard Organization: In ISO/TS 27687:2008. Switzerland, Geneva (2008)
17. Gorria, P., Sevilla, M., Blanco, J.A., Fuertes, A.B.: Carbon 44(10), 1954–1957 (2006)
18. Jin, J., Li, R., Wang, H.L., Chen, H.N., Liang, K., Ma, J.T.: Chem. Commun. 4, 386–388 (2007)
19. Lu, A.H., Schmidt, W., Matoussevitch, N., Bonnemann, H., Spliethoff, B., Tesche, B., Bill, E., Kiefer, W., Schuth, F.: Angewandte Chemie Int Ed. 43(33), 4303–4306 (2004)
20. Park, I.S., Choi, M., Kim, T.W., Ryoo, R.: J. Mater. Chem. 16(33), 3409–3416 (2006)
21. Yang, N., Zhu, S.M., Zhang, D., Xu, S.: Mater. Lett. 62(4–5), 645–647 (2008)
22. Zlotea, C., Chevalier-Cesar, C., Leonel, E., Leroy, E., Cuevas, F., Dibandjo, P., Vix-Guterl, C., Martens, T., Latroche, M.: Faraday Discuss. 151, 117–131 (2011)
23. Tarasov, K., Beaunier, P., Che, M., Marceau, E., Li, Y.L.: J. Nanopart. Res. 13(5), 1873–1887 (2011)
24. Carreno, N.L.V., Garcia, I.T.S., Raubach, C.W., Krolow, M., Santos, C.C.G., Probst, L.F.D., Fajardo, H.V.: J. Power Sources 188(2), 527–531 (2009)
25. Ao, Y., Xu, J., Fu, D., Shen, X., Yuan, C.: Sep. Purif. Technol. 61(3), 436–441 (2008)
26. Carreno, N.L.V., Garcia, I.T.S., Leite, E.R., Longo, E., Lucena, P.R., Carreno, L., Barreto, L.S., Santos, R.: Mater. Lett. 61(16), 3341–3344 (2007)
27. Seo, W.S., Lee, J.H., Sun, X.M., Suzuki, Y., Mann, D., Liu, Z., Terashima, M., Yang, P.C., McConnell, M.V., Nishimura, D.G., Dai, H.J.: Nat. Mater. 5(12), 971–976 (2006)
28. Sharma, A., Nakagawa, H., Miura, K.: Carbon 44(10), 2089–2091 (2006)
29. Zhai, Y., Dou, Y., Liu, X., Park, S.S., Ha, C.-S., Zhao, D.: Carbon 49(2), 545–555 (2010)
30. Ehrburger-Dolle, F., Morfin, I., Geissler, E., Bley, F., Livet, F., Vix-Guterl, C., Saadallah, S., Parmentier, J., Reda, M., Patarin, J., Iliescu, M., Werckmann, J.: Langmuir 19(10), 4303–4308 (2003)
31. Ryoo, R., Joo, S.H., Jun, S.: J. Phys. Chem. B 103(37), 7743–7746 (1999)
32. Vix-Guterl, C., Boulard, S., Parmentier, J., Werckmann, J., Patarin, J.: Chem. Lett. 10, 1062–1063 (2002)
33. Leite, E.R., Carreno, N.L.V., Longo, E., Pontes, F.M., Barison, A., Ferreira, A.G., Maniette, Y., Varela, J.A.: Chem. Mater. 14(9), 3722–3729 (2002)

34. Figuerola, A., Di Corato, R., Manna, L., Pellegrino, T.: Pharmacol. Res. **62**(2), 126–143 (2010)
35. Benjamin, J.S.V., Volin, T.E.: Met. Trans. **5**, 1929–1934 (1974)
36. Rahim, A., Barros, S.B.A., Arenas, L.T., Gushikem, Y.: Electrochim. Acta **56**(3), 1256–1261 (2011)
37. Tavasoli, A., Sadagiani, K., Khorashe, F., Seifkordi, A.A., Rohaniab, A.A., Nakhaeipour, A.: Fuel Process. Technol. **89**(5), 491–498 (2008)
38. Li, X.L., Tian, X.L., Zhang, D.W., Chen, X.Y., Liu, D.J.: Materials science and engineering B-advanced functional. Solid-State Mater **151**(3), 220–223 (2008)
39. Sunny, V., Kumar, D.S., Mohanan, P., Anantharaman, M.R.: Mater. Lett. **64**(10), 1130–1132 (2010)
40. Cava, S., Beninca, R., Tebcherani, S.M., Souza, I.A., Paskocimas, C.A., Longo, E., Varela, J.A.: J. Sol–Gel. Sci. Technol. **43**(1), 131–136 (2007)
41. Carreno, N.L.V., Leite, E.R., Longo, E., Lisboa, P.N., Valentini, A., Probst, L.F.D., Schreiner, W.H.: J. Nanosci. Nanotechnol. **2**(5), 491–494 (2002)
42. Carreno, N.L.V., Maciel, A.P., Leite, E.R., Lisboa, P.N., Longo, E., Valentini, A., Probst, L.F.D., Paiva-Santos, C.O., Schreiner, W.H.: Sensors and Actuators B-Chemical **86**(2–3), 185–192 (2002)
43. Koch, C.C.: Nanostruct. Mater. **9**(1–8), 13–22 (1997)
44. Subrt, J., Perez-Maqueda, L.A., Criado, J.M., Real, C., Bohacek, J., Vecernikova, E.: J. Am. Ceram. Soc. **83**(2), 294–298 (2000)
45. Benvenutti, E.V., Moro, C.C.M., Costa, T.M.H., Gallas, M.R.: Quím. Nova **32** (7), 1926-1933 (2009)
46. Chen, Q., Boothroyd, C., Soutar, A.M., Zeng, X.T.: J. Sol-Gel. Sci. Technol. **53**(1), 115–120 (2010)
47. Ocana, M., Hsu, W.P., Matijevic, E.: Langmuir **7**(12), 2911–2916 (1991)
48. Laranjo, M., Kist, T., Benvenutti, E., Gallas, M., Costa, T.: J. Nanopart. Res. **13**(10), 4987–4995 (2011)
49. Daoush, W.M.: Powder Metall. Met. Ceram. **47**(9–10), 531–537 (2008)
50. Caturla, F., Molina, F., MolinaSabio, M., RodriguezReinoso, F., Esteban, A.: J. Electrochem. Soc. **142**(12), 4084–4090 (1995)
51. Dujardin, E., Ebbesen, T.W., Hiura, H., Tanigaki, K.: Science **265**(5180), 1850–1852 (1994)
52. Moustafa, S.F., Daoush, W.M.: J. Mater. Process. Technol. **181**(1–3), 59–63 (2007)
53. Bao, Q., Zhang, D., Qi, P.: J. Colloid Interface Sci. **360**(2), 463–470 (2011)
54. Alagiri, M., Muthamizhchelvan, C., Ponnusamy, S.: Mater. Lett. **65** (11), 1565–1568
55. Steen, V., Prinsloo, E.F.F.: Catalysis Today **71** (3–4), 327–334 (2002)
56. Tavasoli, A., Sadaghiani, K., Nakhaeipour, A., Ahangari, M.: Iran. J. Chem. Chem. Eng. Int. Engl Ed **26**(1), 9–16 (2007)
57. Ogunniyi, D.S.: Bioresour. Technol. **97**(9), 1086–1091 (2006)
58. Wei, C.Y., Srivastava, D., Cho, K.J.: Nano Lett. **2**(6), 647–650 (2002)
59. Yadav, S., Kumar, A., Pundir, C.S.: Anal. Biochem. **419** (2), 277–283
60. Sunny, V., Sakthi Kumar, D., Yoshida, Y., Makarewicz, M., TabiÅ, W., Anantharaman, M.R.: Carbon **48** (5), 1643–1651 (2010)
61. Hamadi, N.K., Swaminathan, S., Chen, X.D.: J. Hazard. Mater. **112**(1–2), 133–141 (2004)
62. Misirli, T., Bicer, I.O., Mahramanlioglu, M.: Fresenius Environ. Bull. **13**(10), 1010–1015 (2004)
63. Oliveira, L.C.A., Rios, R., Fabris, J.D., Garg, V., Sapag, K., Lago, R.M.: Carbon **40**(12), 2177–2183 (2002)
64. Xing, W., Zhuo, S.P., Gao, X.L.: Mater. Lett. **63**(13–14), 1177–1179 (2009)
65. Zahoor, M., Mahramanlioglu, M.: Chem. Biochem. Eng. Q. **25**(1), 55–63 (2011)
66. Widder, K.J., Senyei, A.E., Scarpelli, D.G.: Proc. Soc. Exp. Biol. Med. **58**, 141–146. (1978)
67. Widder, K.J., Morris, R.M., Poore, G.A., Howard, D.P., Senyei, A.E.: Eur. J. Cancer Clin Oncol. **19**(1), 135–139 (1983)

68. Rudge, S.R., Kurtz, T.L., Vessely, C.R., Catterall, L.G., Williamson, D.L.: Biomaterials **21**(14), 1411–1420 (2000)
69. Lin, Y., Taylor, S., Li, H.P., Fernando, K.A.S., Qu, L.W., Wang, W., Gu, L.R., Zhou, B., Sun, Y.P.: J. Mater. Chem. **14**(4), 527–541 (2004)
70. Koerner, H., Price, G., Pearce, N.A., Alexander, M., Vaia, R.A.: Nat. Mater. **3**(2), 115–120 (2004)
71. Sen, R., Zhao, B., Perea, D., Itkis, M.E., Hu, H., Love, J., Bekyarova, E., Haddon, R.C.: Nano Lett. **4**(3), 459–464 (2004)
72. Grunlan, J.C., Mehrabi, A.R., Bannon, M.V., Bahr, J.L.: Adv. Mater. **16** (2), 150–153, (2004)
73. Supronowicz, P.R., Ajayan, P.M., Ullmann, K.R., Arulanandam, B.P., Metzger, D.W., Bizios, R.: J. Biomed. Mater. Res. **59**(3), 499–506 (2002)
74. Firme, C.P., Bandaru, P.R.: Nanomed. Nanotechnol. Biol. Med. **6**(2), 245–256 (2010)
75. Lu, Q., Moore, J.M., Huang, G., Mount, A.S., Rao, A.M., Larcom, L.L., Ke, P.C.: Nano Lett. **4**(12), 2473–2477 (2004)
76. Sayes, C.M., Liang, F., Hudson, J.L., Mendez, J., Guo, W.H., Beach, J.M., Moore, V.C., Doyle, C.D., West, J.L., Billups, W.E., Ausman, K.D., Colvin, V.L.: Toxicol. Lett. **161**(2), 135–142 (2006)
77. Zhang, L.W., Zeng, L.L., Barron, A.R., Monteiro-Riviere, N.A.: Int. J. Toxicol **26**(2), 103–113 (2007)
78. Hou, P.X., Liu, C., Cheng, H.M.: Carbon **46**(15), 2003–2025 (2008)
79. Ismail, A.F., Goh, P.S., Tee, J.C., Sanip, S.M., Aziz, M.: NANO **3**(3), 127–143 (2008)
80. Tasis, D., Tagmatarchis, N., Bianco, A., Prato, M.: Chem. Rev. **106**(3), 1105–1136 (2006)
81. Trojanowicz, M.: Trac-Trends Anal. Chem. **25**(5), 480–489 (2006)
82. Hurt, R.H., Monthioux, M., Kane, A.: Carbon **44**(6), 1028–1033 (2006)
83. Liu, X., Guo, L., Morris, D., Kane, A.B., Hurt, R.H.: Carbon **46**(3), 489–500 (2008)
84. Braun, T., Rausch, H., Biro, L.P., Konya, L., Kiricsi, I.: J. Radioanal. Nucl. Chem. **262**(1), 31–34 (2004)
85. Banks, C.E., Crossley, A., Salter, C., Wilkins, S.J., Compton, R.G.: Angewandte Chemie-Int. Ed. **45**(16), 2533–2537 (2006)
86. Pumera, M.: Langmuir **23**(11), 6453–6458 (2007)
87. Mello, P.A., Rodrigues, L.F., Nunes, M.A.G., Mattos, J.C.P., Muller, E.I., Dressler, V.L., Flores, E.M.M.: J. Braz. Chem. Soc. **22**(6), 1040–1062 (2011)
88. Mortari, S.R., Cocco, C.R., Bartz, F.R., Dresssler, V.L., Flores, E.M.D.: Anal. Chem. **82**(10), 4298–4303 (2010)
89. Pereira, J.S.F., Antes, F.G., Diehl, L.O., Knorr, C.L., Mortari, S.R., Dressler, V.L., Flores, E.M.M.: J. Anal. At. Spectrom. **25**(8), 1268–1274 (2010)

Performance of Ni/MgAl$_2$O$_4$ Catalyst Obtained by a Metal-Chitosan Complex Method in Methane Decomposition Reaction with Production of Carbon Nanotubes

G. B. Nuernberg, L. F. D. Probst, M. A. Moreira and C. E. M. Campos

Abstract This paper describes the synthesis of Ni/MgAl$_2$O$_4$ catalysts using a method developed by our group with the objective of obtaining a material with more homogeneous composition, more porous structure and greater surface area compared with other spinel preparation methods. The performance of the material obtained was evaluated in the catalytic decomposition of methane, which is a potential alternative route for obtaining pure hydrogen and valuable carbonaceous materials. The textural properties of the catalyst were investigated by X-ray diffraction (XRD), N$_2$ adsorption/desorption isotherms (BET and BJH methods), and temperature-programmed reduction (TPR) analysis. The nature of the carbon deposits was investigated by thermogravimetric analysis (TGA), Raman spectroscopy, scanning electron microscopy (SEM) and transmission electron microscopy (TEM). The influence of the operating conditions on the characteristics of the carbon deposited was studied. The results demonstrated the efficiency of the catalyst in this reaction with the formation of CNTs, irrespective of the operating conditions employed. In general, multiple-walled nanotubes (MWCNTs) were preferentially obtained, and when a diluted flow of CH$_4$ was used the CNTs presented a greater degree of graphitization.

1 Introduction

The use of magnesium aluminate spinel (MgAl$_2$O$_4$) as a support for metallic catalysts is a relatively new application that has provided good results due to some

---

G. B. Nuernberg (✉) · L. F. D. Probst · C. E. M. Campos
CFM, Universidade Federal de Santa Catarina, Florianópolis-SC, 88040-900, Brazil

M. A. Moreira
CAV, Universidade do Estado de Santa Catarina, Lages-SC, 88520-000, Brazil

C. Avellaneda (ed.), *NanoCarbon 2011*, Carbon Nanostructures,
DOI: 10.1007/978-3-642-31960-0_3, © Springer-Verlag Berlin Heidelberg 2013

attractive properties, such as low acidity, hydrophobic character, high thermal resistance and good interaction with the metallic phase, which are of interest for catalytic purposes [1–3]. Conventionally, $MgAl_2O_4$ spinel is prepared using MgO and Al_2O_3 as starting materials and temperatures as high as 1400–1600 °C [2, 3]. However, lower temperatures can be employed in the spinel preparation. The synthetic procedure chosen to prepare the spinel was based on the formation of a metal-chitosan complex, since no report was found in the literature using this preparation technique. This method was developed by our research group—LABOCATH (Laboratório de Catálise Heterogênea), in order to obtain microspheres of Al_2O_3 with high specific surface area [4]. In this case, a hybrid sphere was obtained by metal-chitosan complexation, beyond the basic coagulation, composed of metallic salts of Mg and Al (nitrates) and the biopolymer chitosan. The spinel phase was formed at temperatures of around 500 °C. The $Ni/MgAl_2O_4$ catalyst was obtained through the conventional method of impregnation of the $MgAl_2O_4$ support.

The Ni supported on spinel $MgAl_2O_4$ was used to promote the methane decomposition reaction Eq. (1).

$$CH_{4(g)} \rightarrow C_{(s)} + 2H_{2(g)} \quad \Delta H°_{298K} = 74.52\,kJ.mol^{-1} \tag{1}$$

Methane decomposition is a moderately endothermic reaction and, in general, only hydrogen is detected as the gaseous product. In addition, the decomposition of methane results in the generation of a very important by-product, nanocarbon materials (carbon nanotubes and nanofibers), which have been gaining considerable attention due to their excellent properties and potential applications [5–7]. In this context, we directed this study toward the production of carbon nanotubes (CNTs).

During the methane decomposition reaction, carbon is continuously deposited on the catalyst and forms graphitic structures. Carbon deposition or coke formation is a known cause of the deactivation of metallic catalysts in hydrocarbon processing reactions. The term 'coke' generally refers to a variety of carbon species present as carbonaceous materials containing polyaromatic structures with high a C:H ratio, and amorphous and graphitic carbon. It has been assumed that the graphitic carbon species covers the metallic surface leading to deactivation and degradation of the granular structure of the catalyst due to its volume expansion [8].

The formation of coke on nickel surfaces has been well studied, although some aspects of this process are not at all clear. It was believed that hydrocarbons dissociate to produce highly reactive monoatomic carbon (Cα). Cα is easily turned into a gas, as shown in Eq. (2), to form carbon monoxide [9, 10].

$$C + 1/2O_2 \rightarrow CO \tag{2}$$

If an excess of Cα is formed or the gasification process takes a long time, the polymerization or rearrangement of Cα is favored and the production of Cβ takes place. Tests have shown that polymeric Cβ is less reactive than Cα, and the gasification of this species is quite slow. This results in a buildup of Cβ at the

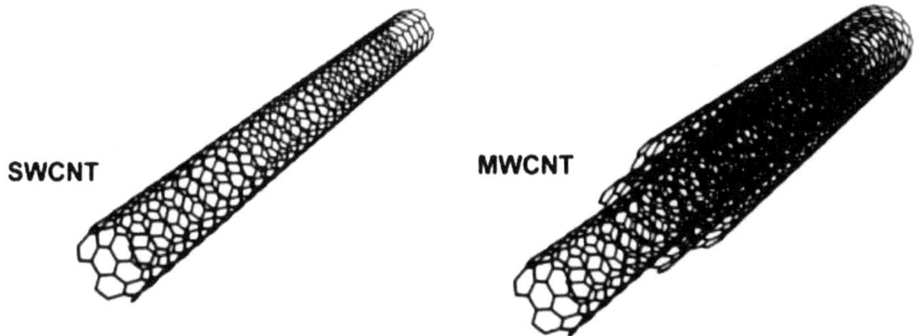

Fig. 1 Single-walled CNT (SWCNT) and multiple-walled CNT (MWCNT). Reprinted with permission from Ref. [12]. Copyright © 2011, American Chemical Society

surface or diffusion in the metal, precipitating at the metal/support interface. This continuous process causes the formation of carbon filaments (as CNTs), which separate the metal from the support surface, initially leading to a greater exposure to active sites. However, the continuous growth causes the filament fragmentation, with loss of the active phase and blocking of the reaction bed. [9].

CNTs have gained considerable interest from researchers due to their diversified structures and remarkable properties (size, chemical stability, thermal conductivity, electronic and mechanical properties, hydrogen storage capability, sensor capability) [10]. The CNTs are comprised solely of carbon and have a well-defined spatial structure, as graphene sheets rolled up as cylinders with nanometric-scale diameters and walls formed of carbon atoms linked in hexagonal arrangements. Depending on the number of graphene layers, CNTs can be classified as: single-walled carbon nanotubes (SWCNT), or multiple-walled carbon nanotubes (MWCNT) (Fig. 1). The SWCNTs are thinner, because they are formed of only one layer of graphene and have diameters varying from 1 to 5 nm. The MWCNTs are comprised of multiples concentric graphene layers, spaced apart from each other by a distance of 0.34 nm and normally have diameters of 10–100 nm with lengths greater than 10 micrometers. Their properties are directed related to the number of layers and inner diameters. The MWNTs are easily experimentally obtained [11].

The structure formation, texture and particular morphology of the carbon have been described in various publications and these characteristics are dependent on the conditions employed in the reaction, such as the catalyst, reaction and reduction temperatures, pressure, reaction feed flow.

The presence of metallic catalysts (such as Fe, Co, Ni) is essential for the formation of CNTs, since these particles act as nucleating agents [8]. In this case, the characteristics of metallic catalysts are directly responsible for the quality of the CNTs formed. Properties such as tube diameter, purity and graphitization grade, among others, are strongly dependent on the type, quality and distribution of the catalyst particles. For instance, the CNT diameter is approximately the same as that of the nanoparticles used in the catalyst [13].

Although advances have been made in recent years, the synthesis of CNTs remains a field with much to be explored, one of the greatest challenges being the acquirement of high purity samples with a homogeneous distribution of diameters, containing only SWCNT or MWCNT, with a high degree of alignment, high yield and fast production rate and low cost [13, 14].

In this context, this paper presents the synthesis of $Ni/MgAl_2O_4$ catalysts, developed by our group, with the objective of obtaining a material with more homogeneous composition, a more porous structure and greater surface area compared with other spinel preparation methods. These catalysts were evaluated in the catalytic decomposition of methane reaction, which is an alternative route for producing carbon nanotubes. The reaction was carried out under different operational conditions and the properties of the CNTs obtained were evaluated as a function of their structure, morphology and quality.

2 Experimental

2.1 Support Preparation

The material used to form the spinel was prepared using a molar ratio of 1.5 chitosan: 1.0 MgO: 1.4 Al_2O_3. In the preparation of the support ($MgAl_2O_4$), 5.0 g chitosan biopolymer [$(C_6H_{11}O_4N)n$] (Purifarma) were dissolved in 167 mL of acetic acid solution (5 % v/v). Concurrently, 21.35 g of $Al(NO_3)_3.9H_2O$ (Vetec) were dissolved in 50 mL of distilled water and 5.21 g of $Mg(NO_3)_2.6H_2O$ (Vetec) were dissolved in 6 mL of distilled water. The Mg and Al aqueous solution was then added to the polymer solution with stirring. The chitosan-Mg–Al solution was added dropwise into an NH_4OH solution (50 % v/v) under vigorous stirring. This resulted in the formation of spheres of chitosan-Mg–Al which were kept in the solution of NH_4OH for approximately 3 h to complete the gel forming process. The spheres were then removed from the basic solution and dried at ambient temperature for 48 h and calcined in an oxidant atmosphere (air) at temperatures between 500 and 1100 °C for 4 h, to form the spinel phase.

2.2 Catalyst Preparation

The Ni addition (20 wt. %) was carried out by impregnation of the material calcined at 1100 °C with an aqueous solution of $Ni(NO_3)_3.6H_2O$ (Sigma-Aldrich), in the appropriate concentrations, with stirring at 80 °C until solvent evaporation. The sample was placed into a drying oven at 90 °C for 16 h and calcined at 700 °C for 5 h in air, and this sample was named 20 %$Ni/MgAl_2O_4E\#700$.

2.3 Catalyst Characterization

The X-ray diffraction (XRD) measurements for the support (MgAl$_2$O$_4$) calcined at 500–1100 °C were taken with a PanAnalytical X'pert PRO Multi-Purpose Diffractometer using Cu Kα radiation (l = 1.5418 Å) operating at 45 kV and 40 mA. The structural characterization obtained from the XRD data was carried out using the GSAS + EXPEGUI program package [15, 16], crystallographic information contained from the ICSD database [17] and the Rietveld method, and with a Siemens D-5000 (Karlsruhe) with Cu Kα irradiation source (λ = 1.540 Å) and a graphite monochromator. The average size of the spinel crystallites was determined through the Scherrer equation.

The specific surface area and pore volume of the catalysts were determined by N$_2$ adsorption/desorption isotherms, at liquid nitrogen temperature, obtained from an Autosorb-1C analyzer (Quantachrome Instruments). The samples were outgassed under vacuum at 200 °C for 2 h. Specific surface areas were calculated according to the Brunauer–Emmett–Teller (BET) method and the pore size distributions were obtained according to the Barret–Joyner–Halenda (BJH) method, from the adsorption data.

Temperature-programmed reduction (TPR) analysis was performed to determine the reducible species present at the surface of the catalyst and the temperature at which these species are reduced by H$_2$ consumption. TPR was carried out on a Micromeritics Chemisorb 2705 analyzer, using 50 mg of catalyst and a temperature ramp from 25 to 1000 °C at 10 °C.min^{-1}. A 30 mL.min^{-1} flow rate of 5 % H$_2$/N$_2$ was used.

The structure, texture and morphology of the deposited carbon were characterized by means of Raman spectroscopy (Renishaw—RGH22) at room temperature, with a 514.5 nm Ar laser.

The thermogravimetric analysis (TGA) was carried out at the Chemistry Department of the Federal University of Santa Catarina (UFSC), Brazil (Shimadzu TGA-50). The analysis was carried out under airflow at 50 mL.min^{-1} and heating from room temperature to approximately 900 °C with a heating rate of 10 °C min^{-1}.

The sample morphology was examined by Scanning Electron Microscopy (SEM) using a Philips XL30 scanning microscope operating at an accelerating voltage of 20 kV and Transmission Electron Microscopy (TEM), with a JEM—1011 microscope operating at an accelerating voltage of 120 kV.

2.4 Catalyst Testing

The decomposition reaction of CH$_4$ was carried out in a quartz-tube fixed-bed flow reactor heated by an electric furnace (Fig. 2). The 20 %Ni/MgAl$_2$O$_4$E#700 catalysts (100 mg) were pre-treated in situ in an H$_2$ stream at 700 °C for 1 h and 2 h,

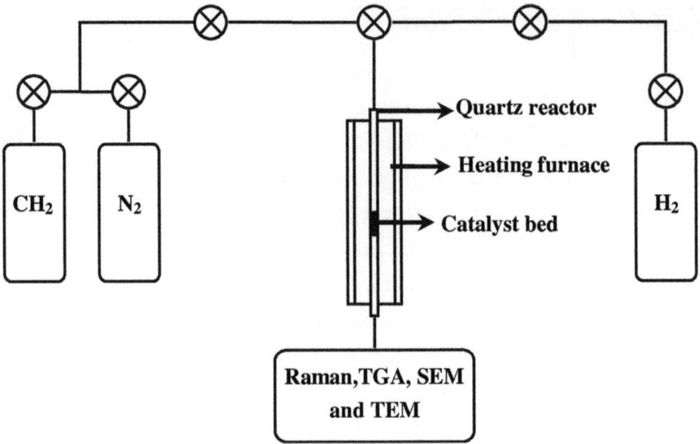

Fig. 2 Schematic diagram of the reactor system

with a heating rate of 10 °C.min^{-1}. The experiments were conducted under atmospheric pressure at 550 °C. The reaction gas was composed of N_2:CH_4 in molar ratios of 7:1; 1:1 and 1:3. The total flow rate of the reaction gas was 80 mL.min^{-1}. The N_2 in the reaction gas was used as a diluent and as an internal analysis standard. The nature of the carbon deposits was investigated as described previously (Raman spectroscopy, TGA, SEM and TEM).

3 Results and Discussions

The determination of the phases present in the materials prepared was carried out by X-ray analysis. Figures 3 and 4 show the X-ray diffractograms of the support calcined at temperatures between 500 and 1100 °C and of the 20 %Ni/MgAl$_2$O$_4$E#700 catalyst, respectively.

Figure 3 shows that the supports prepared through the formation of a metal-chitosan complex followed by calcination contained the spinel phase ($MgAl_2O_4$). This simple ascertainment reveals a great advantage of this preparation method, since the spinel was obtained at a lower calcination temperature (500 °C) than that reported in literature (above 700 °C). However, a rapid quantification indicated that the sample calcined the 1100 °C comprised 68 % $MgAl_2O_4$ and 32 % Al_2O_3. Furthermore, this increase in the calcination temperature promoted an increase in the intensity of the spinel phase peaks, suggesting better quality crystallites [1, 2, 18].

The results for the X-ray diffraction analysis of the 20 %Ni/MgAl$_2$O$_4$E#700 catalyst are shown in Fig. 4. According to the diffraction patterns, the catalyst contains NiO, $MgAl_2O_4$ and/or $NiAl_2O_4$ phases. It is important to note that $NiAl_2O_4$ and $MgAl_2O_4$ spinel-like phases are indistinguishable in XRD analysis, and a more complete characterization can be obtained from the TPR results [19]. The sizes of the crystallites in this sample were calculated using the relation

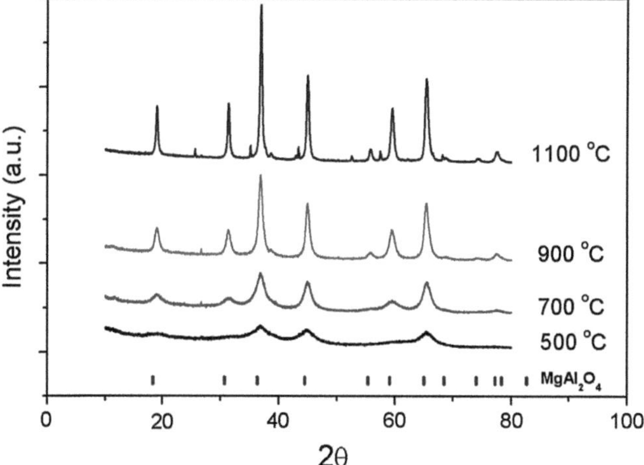

Fig. 3 X-ray diffraction patterns for the support calcined at temperatures between 500 and 1100 °C

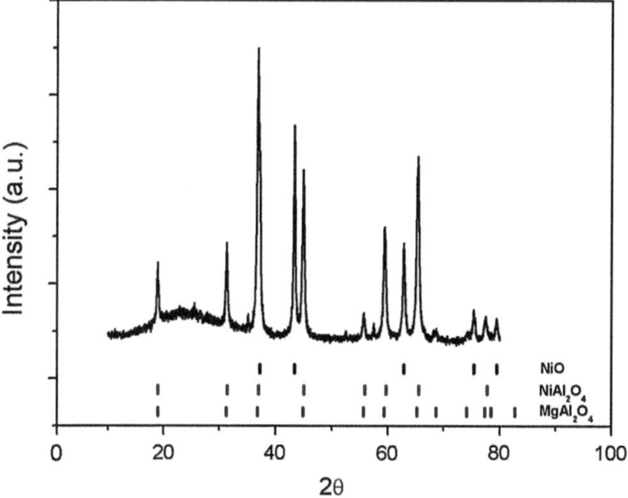

Fig. 4 X-ray diffraction patterns for the 20 %Ni/MgAl$_2$O$_4$E#700 catalyst

between the diffraction patterns and the phases which correspond to the peaks of greatest intensity. It was observed that the crystallite size for NiO was 214 Å and for MgAl$_2$O$_4$ it was 229 Å.

The specific surface area (BET), pore size distribution (BJH) and average pore diameter values of the MgAl$_2$O$_4$ support calcined at temperatures of 500 and 1100 °C (MgAl$_2$O$_4$E#500 and MgAl$_2$O$_4$E#1100, respectively) and of the 20 %Ni/ MgAl$_2$O$_4$E#700 catalyst are summarized in Table 1.

Table 1 Surface properties measured by N_2 physisorption

Samples	Surface Area (m^2g^{-1})	Pore Volume (cm^3g^{-1})	ϕ_m (Å)
$MgAl_2O_4E\#500$	280.31	0.411	58
$MgAl_2O_4E\#1100$	36.22	0.176	194
20 %$Ni/MgAl_2O_4E\#700$	28.16	0.125	177

[a] ϕ_m Average pore diameter

As can be observed in Table 1, the values for the superficial areas as well as the pore volume decrease significantly with an increase in the calcination temperature of the sample and with the impregnation of the metal. This reduction in the superficial area and pore volume of the support calcined at 1100 °C, when compared with that calcined at 500 °C, may be due to sintering of the material at high temperatures. When compared with the 20 %Ni catalyst, this reduction can be related to the presence of Ni species in the interior of the pores, as well as the growth of these particles on the surface of the support, due to the high temperature of calcination.

Figure 5 shows the N_2 adsorption/desorption isotherms for the $MgAl_2O_4E$ support calcined at 500 and 1100 °C and for the 20 %$Ni/MgAl_2O_4E\#700$ catalyst.

The isotherms presented in Fig. 5, according to IUPAC, are of type V for the $MgAl_2O_4E\#1100$ and 20 %$Ni/MgAl_2O_4E\#700$ samples, characteristic of a solid with mesopores, with exception of the $MgAl_2O_4E\#500$ sample, where the isotherm is of type II, characteristic of non-porous solids or those with macropores [20, 21].

Figure 6 shows the pore size distributions for the same samples.

The curves for the pore volume distribution (Fig. 6) and the average diameter of the pores (Table 1), show the formation of pores with diameters ranging between 58 and 218 Å for the three samples, a characteristic predominantly of mesoporous materials (20–500 Å).

Temperature-programmed reduction (TPR) analysis was performed in order to determine the reducible species at the surface of the 20 %$Ni/MgAl_2O_4E\#700$ catalyst and the temperature at which these species are reduced (Fig. 7).

Figure 7 shows the TPR curve for the 20 %$Ni/MgAl_2O_4E\#700$ sample where three peaks can be observed with maximums at around 400, 700 and 900 °C. The first peak can be attributed to a weak interaction of the NiO reduction with $MgAl_2O_4$, the second peak to a strong interaction of the Ni reduction with $MgAl_2O_4$ and the third to the reduction of the spinel (Ni, Mg) Al_2O_4 [19, 22, 23]. Similar behavior was observed by Park et al. [23], for a $Ni/MgAl_2O_4$ catalyst calcined at 700 °C in an oxidant atmosphere. According to the author, a general AB_2O_4 spinel structure consists of A^{2+} and B^{3+} ions where the cations can easily displace each other. Therefore, Mg^{2+} in $MgAl_2O_4$ could substitute Ni^{2+} upon heat treatment at 700 °C in air. $NiAl_2O_4$ spinel is created during these processes. The authors also observed that when $Ni/MgAl_2O_4$ was heat treated in H_2, some Ni^{2+} ions were resubstituted by Mg^{2+}, and converted to surface Ni^{2+}.

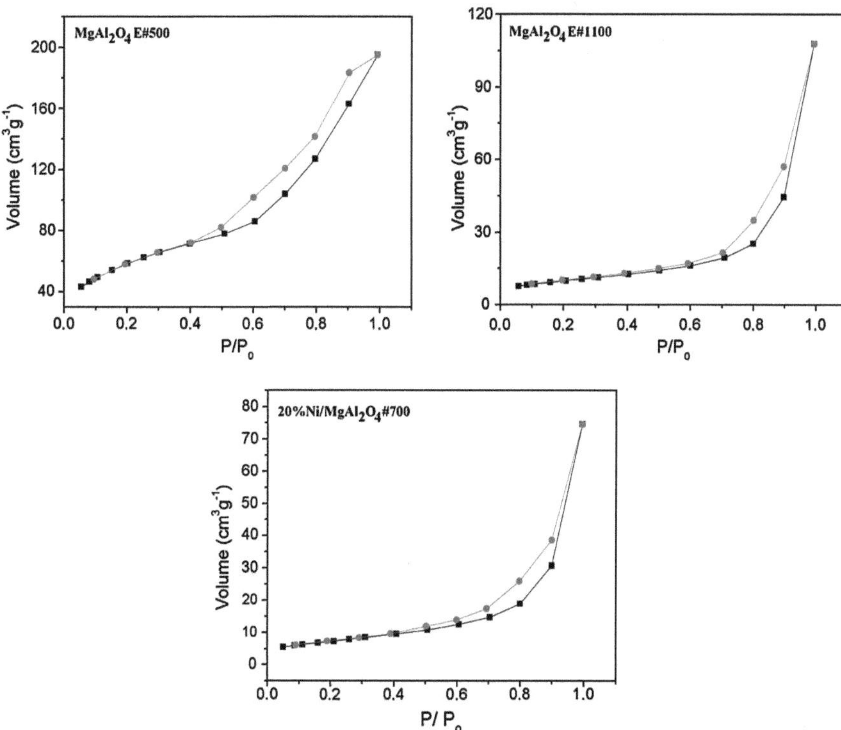

Fig. 5 N$_2$ adsorption/desorption isotherms: (■) adsorption and (●) desorption curves of the support calcined at 500 and 1100 °C (MgAl$_2$O$_4$E#500 and MgAl$_2$O$_4$E#1100, respectively) and of the 20 %Ni/MgAl$_2$O$_4$#700 catalyst

The SEM images in Fig. 8 show the 20 %Ni/MgAl$_2$O$_4$E#700 catalyst surfaces before the catalytic test.

It can be observed that the surface of the catalyst (20 %-Ni/MgAl$_2$O$_4$E#700) with the support prepared through the formation of a metal-chitosan complex presented a porous appearance, with some clusters. In Fig. 8, a smoother appearance can also be noted, due to the high temperatures used for the calcination of the support (1100 °C) and the catalyst (700 °C). These results verify that the aim of obtaining porous materials was achieved, despite the high temperatures involved in the process. These results are in agreement with those of the N$_2$ physisorption analysis (Table 1) reported above.

After the catalytic tests, the solid carbon deposit was observed on the surface of the catalyst. The different structures of this deposited carbon, as a function of the operational conditions employed, were analyzed by Raman spectroscopy, TGA, SEM and TEM.

The Raman spectra of the catalysts after the catalytic tests are shown in Fig. 9 and show bands originating from carbon structures.

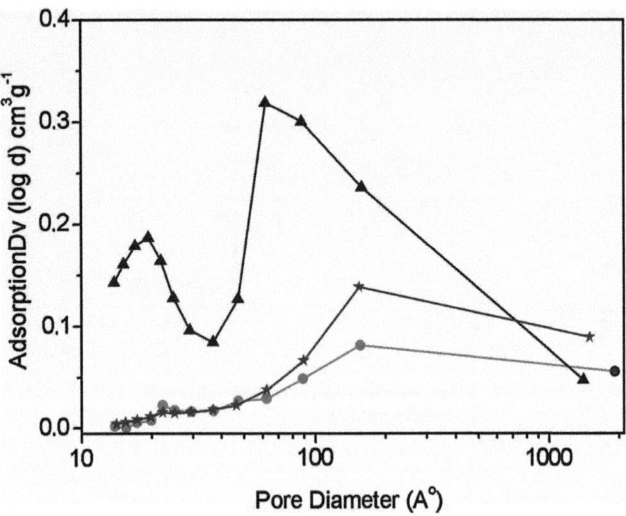

Fig. 6 Pore diameter distribution calculated using the BJH model from the adsorption curve for the support calcined at 500 °C (▲), 1100 °C (★) and the 20 %Ni/MgAl$_2$O$_4$#700 catalyst (●)

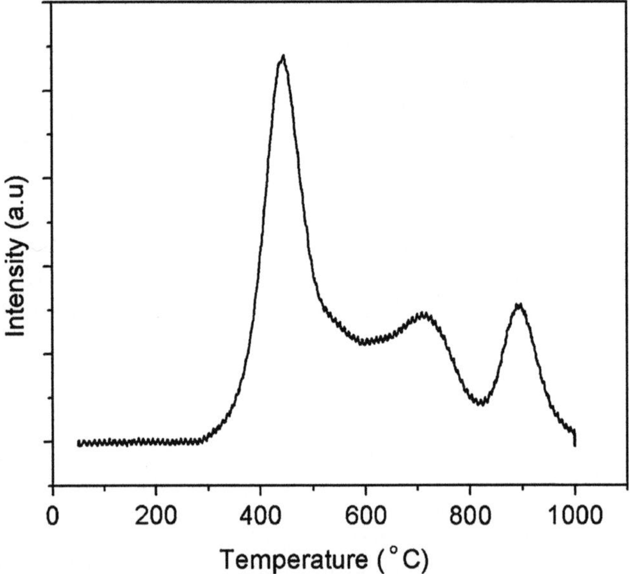

Fig. 7 TPR profile of the catalyst 20 %Ni/MgAl$_2$O$_4$E#700

The Raman spectra of the carbon nanotubes show peaks characteristic of the G-band at 1575 cm^{-1}, attributed to the tangential C–C stretching, and of the D-band at 1350 cm^{-1}, attributed to disordered carbon structures present in carbon

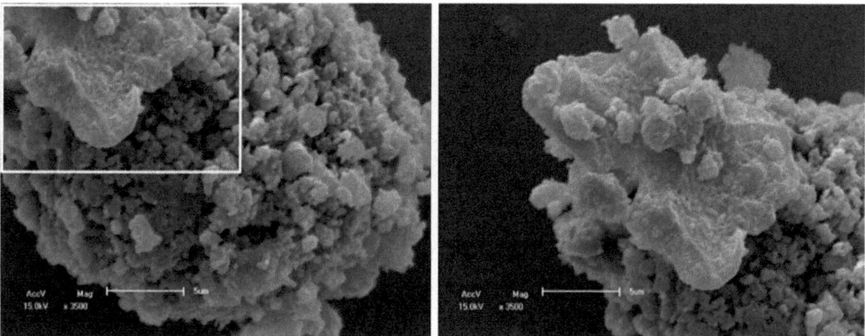

Fig. 8 SEM images of the 20 %-Ni/MgAl$_2$O$_4$E#700 catalyst surfaces before the catalytic test

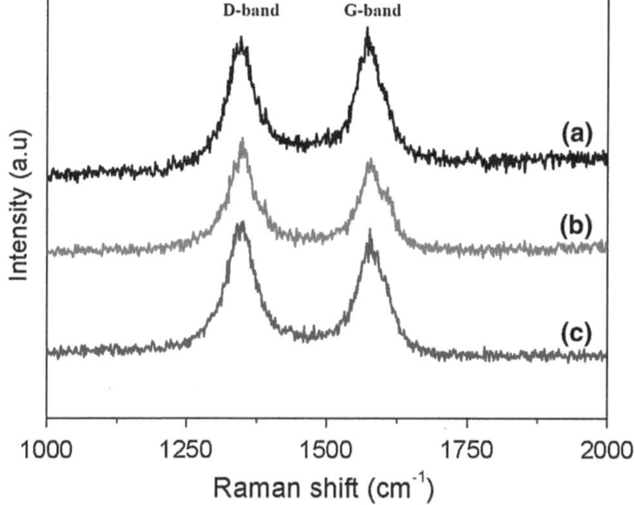

Fig. 9 Raman spectra of the 20 %Ni/MgAl$_2$O$_4$E#700 sample after the catalytic decomposition of methane, with reaction temperature of 550 °C and reduction at 700 °C for 1 h. Molar ratio (N$_2$:CH$_4$): **a** 7:1, **b** 1:1 and **c** 1:3

nanotubes or other carbon forms. The appearance of the D-band is a clear indication of the formation of multiwalled nanotubes (MWNTs). Furthermore, the I_D/I_G ratio can provide important information regarding the quality of carbon nanotubes. Based on the Raman spectrum in Fig. 9, the I_D/I_G ratios can be estimated to be around (a) 0.91, (b) 1.85 and (c) 1.07. A lower I_D/I_G ratio indicates a higher degree of graphitization of the carbon nanotubes. Thus, for the catalyst under this reaction condition, (a) produced the highest degree of graphitization of the carbon nanotubes [24–27].

TGA is a very powerful technique for determining the quality of the synthesized CNTs due to differences in the oxidation stability and degree of

Table 2 Data on percentage of weight loss and temperature of maximum rate of decomposition of the 20 %Ni/MgAl$_2$O$_4$E#700 sample, after catalytic decomposition of the methane as a function of the operational conditions

Entry	Operational Conditions			TGA	
	N$_2$:CH$_4$	T$_R$[a] (°C)	T$^a_{Red}$ (°C)/time (h)	Mass loss (%)	Temperature (°C)
1	1:3	550	700/1	65	603
2	1:1	550	700/1	57	589
3	7:1	550	700/1	43	607

[a] T_R reaction temperature; T_{Red} reduction temperature

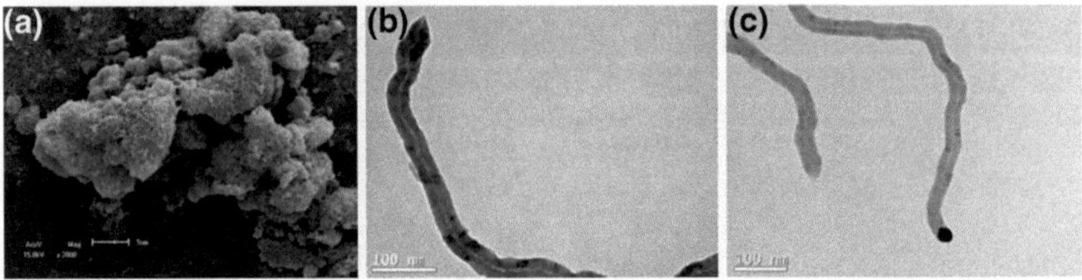

Fig. 10 SEM image of the 20 %-NiMgAl$_2$O$_4$E#700 catalyst surfaces after the catalytic tests, at reaction temperature of 550 °C and reduction at 700 °C for 1 h. Molar ratio: **a** N$_2$:CH$_4$ = 7:1. TEM images for the same catalyst and operational conditions. Molar ratio: **b** N$_2$:CH$_4$ = 7:1 and **c** N$_2$:CH$_4$ = 1:3

graphitization. Different types of carbon, such as amorphous carbon and CNT, appear to have different oxidation temperatures and thus can be distinguished using the TGA technique [28]. Table 2 shows the data for the percentage of weight loss and the temperature of the maximum decomposition rate for the 20 %Ni/MgAl$_2$O$_4$E#700 sample, obtained by TGA analysis, after the catalytic tests.

The thermogravimetric analysis of the 20 %Ni/MgAl$_2$O$_4$E#700 catalyst after the catalytic tests with N$_2$:CH$_4$ molar ratios of 1:3 (entry 1), 1:1 (entry 2) and 7:1 (entry 3) showed a weight loss in the region of 589–607 °C. It can be observed for entries 1 and 3 that there was a weight loss of 65 % at 603 °C and 43 % at 607 °C, respectively. The mass loss of this material at above 600 °C can be attributed to the MWNTs. In the case of entry 2 there was a weight loss of 57 % at 589 °C. This may be associated with small amounts of amorphous particles in the nanotubes, the influence of metal particles and/or defects on the surface of the nanotubes. These factors can affect the temperature at which the maximum decomposition rate occurs. It can be concluded that irrespective of the operational conditions employed a more stable material (CNT) was obtained [14, 29–34].

The results for the SEM and TEM characterization of the carbon-containing materials after the CH$_4$ reaction are shown in Fig. 10.

The images obtained from the SEM (Fig. 10a) and TEM (Fig. 10b and c) analyses show that after the catalytic tests there were modifications on the catalyst surface due to the deposition of carbonaceous material. The images show that the formation of CNTs occurred irrespective of the operational conditions. The TEM images in Fig. 10 show carbon nanotubes with external diameters below 33 nm, which is in agreement with the results obtained in the Raman spectroscopy (Fig. 9) and TGA (Table 2). The results show that there was the preferential formation of carbon nanotubes in the all cases, after the decomposition reaction.

Some black points can be seen in the images of Figs. 10b and (c), which may be attributed to the amorphous carbon and nanoparticles of Ni, respectively, which were formed and deposited on the walls of the carbon nanotubes. Similar behavior has been described by Hsieh et al. [34], Zhou et al. [35] and Guevara et al. [36], employing metal and bimetal catalysts of Pt, Pt–Ni (Fe, Co) and Ni-Ce, respectively.

4 Conclusions

The results indicated that in the preparation of the support via a metal-chitosan complex the spinel phase (MgAl$_2$O$_{-4}$) was obtained at a lower calcination temperature (500 °C) than those reported in the literature ($\geq$700 °C). The 20 %Ni/ MgAl$_2$O$_4$#700 catalyst is composed of a MgAl$_2$O$_4$ spinel phase and/or NiAl$_2$O$_4$ spinel phase (which are indistinguishable in XRD analysis) and an NiO phase. The results demonstrated the efficiency of the 20 %Ni/MgAl$_2$O$_4$ catalyst when applied in the decomposition of CH$_4$ with the formation of carbon nanotubes being verified, independent of the N$_2$:CH$_4$ molar ratio employed. In general, the preferential formation of multiwalled nanotubes (MWCNTs) was observed, and when a diluted flow of CH$_4$ was used the CNTs presented a greater degree of graphitization (I$_D$/I$_G$ = 0.91).

Acknowledgments The authors are grateful to Universidade Federal de Santa Catarina (UFSC) for access to facilities including LCME, LDRX and LabMat, and to the Brazilian government agency Conselho Nacional de Desenvolvimento Científico e Tecnológico (CNPq) for financial support.

References

1. Bocanegra, S.A., Ballarini, A.D., Scelza, O.A., Miguel, S.R.: The influence of the synthesis routes of MgAl$_2$O$_4$ on its properties and behavior as support of dehydrogenation catalysts. Mater. Chem. Phys. **111**, 534–541 (2008). doi:10.1016/j.matchemphys.2008.05.002
2. Folleto, E., Alves, R.W., Jahn, S.L.: Preparation of Ni/Pt catalysts supported on spinel (MgAl$_2$O$_4$) for methane reforming. J. Power Sour. **161**, 531–534 (2006). doi:10.1016/ j.jpowsour.2006.04.121
3. Kong, L.B., Ma, J., Huang, H.: MgAl$_2$O$_4$ spinel phase derived from oxide mixture activated by a high-energy ball milling process. Mater. Lett. **56**, 238–243 (2002). doi:10.1016/S0167-577X(02)00447-0

4. Nuernberg, G.B., Fajardo, H.V., Mezalira, D.Z., Casarin, T.J., Probst, L.F.D., Carreño, N.L.V.: Preparation and evaluation of Co/Al_2O_3 catalysts in the production of hydrogen from thermo-catalytic decomposition of methane: Influence of operating conditions on catalyst performance. Fuel **87**, 1698–1704 (2008). doi:10.1016/j.fuel.2007.08.005
5. Abbas, H.F., Wan Daud, W.M.A.: Hydrogen production by methane decomposition: A review. Int. J. Hydrogen Energy **35**, 1160–1190 (2010). doi:10.1016/j.ijhydene.2009.11.036
6. Botas, J.A., Serrano, D.P., Guil-López, R., Pizarro, P., Gómez, G.: Methane catalytic decomposition over ordered mesoporous carbons: A promising route for hydrogen production. Int. J. Hydrogen Energy **35**, 9788–9794 (2010). doi:10.1016/j.ijhydene.2009.10.031
7. Pinilla, J.L., Suelves, I., Lázaro, M.J., Moliner, R., Palacios, J.M.: Parametric study of the decomposition of methane using a $NiCu/Al_2O_3$ catalyst in a fluidized bed reactor. Int. J. Hydrogen Energy **35**, 9801–9809 (2010). doi:10.1016/j.ijhydene.2009.10.008
8. Li, Y., Li, D., WANG, G.: Methane decomposition to COx-free hydrogen and nano-carbon material on group 8–10 base metal catalysts: A review. Catal. Today **162**, 1–48 (2011). doi:10.1016/j.cattod.2010.12.042
9. Figueiredo, J.L., Ribeiro, F.R.: Catálise Heterogênea. Fundação Calouste Gulbenkian, Lisboa (1989)
10. Norinaga, K., Huttinger, K.J.: Kinetics of surface reactions in carbon deposition from light hydrocarbons. Carbon **41**, 1509–1514 (2003). doi:10.1016/S0008-6223(03)00097-6
11. Paschoalino, M.P., Marcone, G.P.S., Jardim, W.F.: Os nanomateriais e a questão ambiental. Quim. Nova **33**, 421–430 (2010). doi:10.1590/S0100-40422010000200033
12. Schnorr, J.M., Swager, T.M.: Emerging applications of carbon nanotubes. Chem. Mater. **23**, 646–657 (2011). doi:10.1021/cm102406h
13. Zarbin, A.J.G.: Química de (nano)materiais. Quim. Nova **30**(6), 1469–1479 (2007). doi:10.1590/S0100-40422007000600016
14. Maccallini, E., Tsoufis, T., Policicchio, A., La Rosa, S., Caruso, T., Chiarello, G., Colavita, E., Formoso, V., Gournis, D., Agostino, R.G.: A spectro-microscopic investigation of Fe–Co bimetallic catalysts supported on MgO for the production of thin carbon nanotubes. Carbon **48**, 3434–3445 (2010). doi:10.1016/j.carbon.2010.05.039
15. Larson, AC., Von Dreele, RB.: General structure analysis system (GSAS), Los Alamos National Laboratory Report LAUR 86–748 (2000)
16. Toby, B.H.: A graphical user interface for GSAS. J. Appl. Cryst. **34**, 210–213 (2001). doi:10.1107/S0021889801002242
17. Inorganic Crystal Structure Database (ICSD): Gmelin-Institut für Anorganische Chemie and Fachinformationszentrum. FIZ, Karlsruhe (2007)
18. Shiono, T., Shiono, K., Miyamoto, K., Pezzotti, G.: Synthesis and characterization of $MgAl_2O_4$ spinel precursor from a heterogeneous Alkoxide solution containing fine MgO powder. J. Am. Ceram. Soc. **83**, 235–237 (2000). doi:10.1111/j.1151-2916.2000.tb01180.x
19. Monzón, A., Latorre, N., Ubieto, T., Royo, C., Romeo, E., Villacampa, J.I., Dussault, L., Dupin, J.C., Guimon, C., Montioux, M.: Improvement of activity and stability of Ni–Mg–Al catalysts by Cu addition during hydrogen production by catalytic decomposition of methane. Catal. Today **116**, 264–270 (2006). doi:10.1016/j.cattod.2006.05.085
20. Teixeira, V.G., Coutinho, F.M.B., Gomes, A.S.: Principais métodos de caracterização da porosidade de resinas à base de divinilbenzeno. Quim. Nova **24**(6), 808–818 (2001). doi:10.1590/S0100-40422001000600019
21. Silva, JB.: Caracterização de materiais catalíticos. Qualificação de Doutorado, Instituto Nacional de Pesquisas Espaciais (INPE). (2008)
22. Ozdemir, H., Oksuzomer, M.A.F., Gurkaynak, M.A.: Preparation and characterization of Ni based catalysts for the catalytic partial oxidation of methane: Effect of support basicity on H_2/CO ratio and carbon deposition. Int. J. Hydrogen Energy **35**, 12147–12160 (2010). doi:10.1016/j.ijhydene.2010.08.091

23. Park, D.S., Li, Z., Devianto, H., Lee, H.: Characteristics of alkali-resistant Ni/MgAl$_2$O$_4$ catalyst for direct internal reforming molten carbonate fuel cell. Int. J. Hydrogen Energy **35**, 5673–5680 (2010). doi:10.1016/j.ijhydene.2010.03.043
24. Zeng, L., Wang, W., Lei, D., Liang, J., Zhao, H., Zhao, J., Kong, X.: High-field electron emission of carbon nanotubes grown on carbon fibers. Phys. B **403**, 2662–2665 (2008). doi:10.1016/j.physb.2008.01.032
25. Herbst, M.H., Macêdo, M.I.F., Rocco, A.M.: Tecnologia dos nanotubos de carbono: tendências e perspectivas de uma área multidisciplinar. Quim. Nova **27**(6), 986–992 (2004). doi:10.1590/S0100-40422004000600025
26. Fu, J., Huang, Y., Pan, Y., Zhu, Y., Huang, X., Tang, X.: An attempt to prepare carbon nanotubes by carbonizing polyphosphazene nanotubes with high carbon content. Mater. Lett. **62**, 4130–4133 (2008). doi:10.1016/j.matlet.2008.06.020
27. Almeida, R.M., Fajardo, H.V., Mezalira, D.Z., Nuernberg, G.B., Noda, L.K., Probst, L.F.D., Carreño, N.L.V.: Preparation and evaluation of porous nickel-alumina spheres as catalyst in the production of hydrogen from decomposition of methane. J. Mol. Catal. A: Chem. **259**, 328–335 (2006). doi:10.1016/j.molcata.2006.07.044
28. Chen, C.M., Dai, Y.M., Huang, J.G., Jehng, J.M.: Intermetallic catalyst for carbon nanotubes (CNTs) growth by thermal chemical vapor deposition method. Carbon **44**, 1808–1820 (2006). doi:10.1016/j.carbon.2005.12.043
29. Musumeci, A., Silva, G., Martens, W., Waclawik, E., Frost, R.: Thermal decomposition and electron microscopy studies of single-walled carbon nanotubes. J. Therm. Anal. Calorim. **88**(3), 885–891 (2007). doi:10.1007/s10973-006-7563-9
30. Ramesh, B.P., Blau, W.J., Tyagi, P.K., Misra, D.S., Ali, N., Gracio, J., Cabral, G., Titus, E.: Thermogravimetric analysis of cobalt-filled carbon nanotubes deposited by chemical vapour deposition. Thin Solid Films **494**, 128–132 (2006). doi:10.1016/j.tsf.2005.08.220
31. Suriani, A.B., Azira, A.A., Nik, S.F., Nor, R.M., Rusop, M.: Synthesis of vertically aligned carbon nanotubes using natural palm oil as carbon precursor. Mater. Lett. **63**, 2704–2706 (2009). doi:10.1016/j.matlet.2009.09.048
32. Das, N., Dalai, A., Mohammadzadeh, J.S.S., Adjaye, J.: The effect of feedstock and process conditions on the synthesis of high purity CNTs from aromatic hydrocarbons. Carbon **44**, 2236–2245 (2006). doi:10.1016/j.carbon.2006.02.040
33. Zarabadi-Poor, P., Badiei, A., Yousefi, A.A., Fahlman, B.D., Abbasi, A.: Catalytic chemical vapour deposition of carbon nanotubes using Fe-doped alumina catalysts. Catal. Today **150**, 100–106 (2010). doi:10.1016/j.cattod.2009.06.019
34. Hsieh, C., Lin, J., Wei, J.: Deposition and electrochemical activity of Pt-based bimetallic nanocatalysts on carbon nanotube electrodes. Int. J. Hydrogen Energy **34**, 685–693 (2009). doi:10.1016/j.ijhydene.2008.11.008
35. Zhou, M., Lin, G., Zhang, H.: Pt Catalyst Supported on Multiwalled Carbon Nanotubes for Hydrogenation-Dearomatization of Toluene. Chin. J. Catal. **28**(3), 210–216 (2007). doi:10.1016/S1872-2067(07)60020-5
36. Guevara, J.C., Wang, J.A., Chen, L.F., Valenzuela, M.A., Salas, P., García-Ruiz, A., Toledo, A., Cortes-Jácome, M.A., Angeles-Chavez, C., Novaro, O.: Ni/Ce-MCM-41 mesostructured catalysts for simultaneous production of hydrogen and nanocarbon via methane decomposition. Int. J. Hydrogen Energy **35**, 3509 (2010). doi:10.1016/j.ijhydene.2010.01.068

The Use of Nanostructures for DNA Transfection

Vinicius Farias Campos, Virgínia Yurgel, Fabiana Kömmling Seixas and Tiago Collares

Abstract The interaction between nanostructured materials and living systems is of fundamental and practical interest and will determine the biocompatibility, potential utilities and applications of novel nanomaterials in biological settings. The pursuit of new types of molecular transporters is an active area of research, due to the high impermeability of cell membranes and other biological barriers to foreign substances and the need for intercellular delivery of molecules via cell-penetrating transporter for drug, gene or protein therapeutics. Here, is described the novel nanostructure-based transfection systems. The transfection uses of nanopolymers, nanoparticles and nanotubes are the main focus of this review. In addition are described the technique called NanoSMGT that uses nanostructures for DNA transfection in sperm cells that could be used for transgenic animal generation or human gene therapy.

1 Introduction

The delivery of genetic materials into target cells is a powerful tool to manipulate or modify the cell physiology and to induce the production of specific proteins by cells and have been called as DNA/RNA transfection. Apart from the new sophisticated transfection systems, naked DNA was the first used on transfection procedures without any vector to their cellular internalization [39]. The precipitation of DNA with divalent cations was used for the transfection of cultured cell

V. F. Campos (✉) · V. Yurgel · F. K. Seixas · T. Collares
Grupo de Pesquisa em Oncologia Celular e Molecular, Programa de Pós-Graduação em Biotecnologia, Centro de Desenvolvimento Tecnológico, Universidade Federal de Pelotas, Pelotas-RS, Brazil
e-mail: fariascampos@gmail.com

C. Avellaneda (ed.), *NanoCarbon 2011*, Carbon Nanostructures, 65
DOI: 10.1007/978-3-642-31960-0_4, © Springer-Verlag Berlin Heidelberg 2013

lines for more than 10 years (calcium phosphate precipitation). However the transfection efficiency using naked DNA remained low.

Several transfection techniques have been developed during the last 30 years. Viruses have evolved formidable solutions to gene transfer. Consequently, genetically modified (recombinant) viruses rank among the most efficient vehicles known today for the transfer of foreign genetic information into cells [35]. A multitude of viral species have been engineered as gene vector, including retroviruses, adenoviruses, adeno-associated viruses, herpes simplex viruses, hepatitis viruses, vaccinia viruses and lentiviruses. In general, the genetic information required for the natural replicative cycle of the virus is removed from the viral genome and replaced by the genes of interest [47]. Although viral transfection has a better efficiency in DNA delivery to the target cell, there are still several problems in its use such as high cytotoxicity, the high cost, low reproducibility and concerns about biosafety [2].

As an alternative to viral gene vectors, non-viral, synthetic and half-synthetic transfection systems have been developed. Most of these non-viral vectors mimic important features of viral cell entry in order to overcome the cellular barriers to infiltration of foreign genetic material [47]. Among these barriers are the plasma membrane, membranes of internal vesicles such as endosomes and lysosmes and the nuclear membranes [3]. Among the functions mimicked in non-viral vectors are the capability of receptor targeting, of DNA binding and compation and of intracellular release from internal vesicles. These individual functions are represented in synthetic and half-synthetic modules which usually are assembled by electrostatic and/or hydrophobic interactions to form a particle vector for transfection procedure [35]. The lipoplexes are assemblies of nucleic acids with a lipidic component, which is usually cationic. Gene transfection by lipoplexes is called lipofection. Lipofection is a method in which exogenous DNA is transported inside cationic liposomes to cross the plasma membrane by endocytosis [16]. However, until the internalized molecules reaching the nucleus, enzymes that degrade DNA called DNases can hydrolyse most of the transfected DNA, reducing the efficiency of this technique [7]. In addition, lipoplexes caused several changes to cells, which included cell shrinking, reduced cell mitoses and cytoplasm vacuolization [31].

In order to enhance the DNA transfection process, new compounds have been tested. Among these, nanostructures have been developed specifically for the transfection procedures. These nanomolecules have shown promising characteristics for the delivery of genetic material. A significant advantage is the nanomeric scale of these structures, which facilitates the entry into the cell and allows high cell viability during the transfection process. In particular, the small size of these nanostructures allows common biological processes of the cell (e. g. endocytosis) are used for internalization [4]. Moreover, these nanostructures protect the DNA associated with them from DNase enzyme action that degrades genetic material during intracellular trafficking. These features make the nanostructures appropriate candidates to carry exogenous DNA/RNA molecules across cell membranes [17, 36].

Several functionalized and non-functionalized nanostructures have been developed for DNA transfection, including nanotubes and magnetic nanotubes, nanoparticles and magnetic nanoparticles, nanopolymers and dendrimers.

2 Nanopolymers and Dendrimers for DNA Transfection

The main goal on use of nanostructures for DNA/RNA transfection is that nano-composites such as nanotubes, nanoparticles or cationic dendrimers can protect DNA strands against nucleases during cellular delivery acting as a physical barrier to DNase enzymes [28].

Today a series of new nanostructures have been developed for DNA transfection being some of them commercially available. Nanostructures such as cationic dendrimers or nanopolimers have been demonstrated to display considerable transfection efficiency both in vivo and in vitro in the past decade [46]. Generally they have abundant primary amines which readily form polyplexes with negatively charged DNA and subsequently buffer the endosomal environment, facilitating the release of DNA in the cytosol [18]. Cationic polymers such as polyethylenimine (PEI), polyamidoamine (PAMAM), polypropylenimine (PPI), poly-L-lysine (PLL), cationic dextran, polyallylamine (PAA), dextran-oligoamine bases conjugates and chitosan are amongst the preferable materials for the preparation of non-viral vectors in terms of their long-term safety and biocompatibility [22].

PLL, PAA and many others were abandoned due to its low transfection efficiency and higher cytotoxicity. Dextran-oligoamine based transfection in a wide range of cell lines is very low in comparison to other nanopolymers based on PEI and dendrimers [22]. Of these, PEI is one of the most successful and widely studied gene delivery polymers due to its membrane destabilization potential, high charge density (DNA condensation capability) and ability to protect endocytosed DNA from enzymatic degradation, thus perform DNA transfer efficiently into the cells [46]. Branched PEI contains primary, secondary and tertiary providing remarkable buffering capacity. The primary amines are mainly responsible for efficient DNA binding ability, but contribute maximum toxicity during transfection, while the secondary and tertiary amino groups provide good buffering capacity to the system [34]. Though high charge density of the system increases the transfection efficiency, concurrently, it contributes to cellular toxicity. Actually, the efficacy of the transfection system is a sweet compromise between the transfection efficiency and the cytotoxicity.

Both linear and branched PEI can be used for transfection in vitro and in vivo. However, it has been reported that linear PEI is less toxic, more efficient and faster acting transfection agent than branched PEI. This could be attributed to the phenomenon that linear PEI-DNA complexes are less condensed and they can penetrate the cell wall and subsequently the cell nucleus more efficiently than branched PEI and DNA complexes [14].

PEI is slightly toxic to the cells and this can be explained by its nonbiodegradable nature. Studies have shown that PEI transfection efficiency is dependent on its molecular weight. The most active are 25 kDa branched PEI and 22 kDa linear PEI [14]. Longer linear PEI also show similar transfection activity, but they are more toxic. On the contrary, shorther linear PEI is less toxic and less efficient.

3 Nanoparticles for DNA Transfection

In general, nanoparticles have been very ineffective vehicles for gene delivery, with expression levels below those seen with naked DNA [5]. Thus, there has been relatively little progress with DNA incorporation into biodegradable sustained release particles. Also, problems encountered in gene therapy include slow accumulation and low concentration of gene vector in target tissues.

Nanoparticles formed from biodegradable polymers have been used to carry active molecules to sites in the body where the therapeutic effect is required. In general, such nanoparticles have limited loading capacity for most hydrophilic drugs and also are not efficient in cases where rapid accumulation of active molecules is required at their target sites [42].

Various studies were conducted to improve delivery of a biomaterial such as an application of a magnetic field to a vector including magnetically responsive solid phases, which are micro-to nanometer sized particles or aggregates.

Magnetic nanoparticles have been primarily applied to three fields: magnetic resonance imaging, molecular and cell separation technologies, and drug delivery [24]. Because these agents are used primarily in diagnostic in vivo imaging, many of the particle formulations are already approved for use in humans. The magnetic properties of these particles are quite favorable for construction of a nonviral-based gene delivery system. Magnetofection is a new method for gene transfer that involves the use of magnetic force and plasmid DNA (pDNA)/magnetic bead complexes, and it has been developed for enhancing delivery of gene vectors to target cells [20]. For magnetofection, pDNA was interacted with magnetic beads and attracted to target cells by magnetic force in order to accumulate on the cell surface.

This method associates nucleic acids or other vectors with magnetic nanoparticles coated with cationic molecules. Generally, the magnetic nanoparticles are made of iron oxide, which is fully biodegradable, coated with specific cationic proprietary molecules varying upon applications. Their association with the gene vectors is achieved by salt-induced colloidal aggregation and electrostatic interaction. The resulting molecular complexes are then targeted to and endocytosed by cells, supported by an appropriate magnetic field. Membrane architecture and structure stay intact in contrast to other physical transfection methods that damage, create hole or electroshock the cell membranes. In addition the magnetic nanobeads are efficiently cleared from the cells and tissues, and are not toxic at the recommended doses and even higher.

The association of magnetic nanoparticles and several other compounds to improve transfection efficiency has been reported in several studies. Ino et al. [23] demonstrated the production of magnetic cationic liposomes (MCLs), containing 10 nm magnetite nanoparticles in order to improve the accumulation of magnetite nanoparticles in target cells through electrostatic interactions between MCLs (positively charged) and the cell membrane (negatively charged). Since MCLs are positively charged, pDNA (negatively charged) interacts with the MCLs electrostatically. This method increases significantly the DNA uptake by cultured cells, however cellular damages were increased. Another magnetofection variation uses self-assembled ternary complexes of cationic magnetic nanoparticles, plasmid DNA and cell-penetrating Tat peptide [42]. Tat is a highly cationic peptide derived from HIV-1 tat protein, with a linear sequence of 13 amino acids. This specific sequence carries a transmembrane signal and a nuclear localisation signal. The membrane absorption is primarily promoted by ionic interaction between the cationic charges of the Tat peptide and anionic charges of the phospholipid heads in biomembranes.

Benefiting from its cell membrane penetrating property, the Tat peptide is capable of mediating intracellular delivery of many different types of cargos [4], including magnetic nanoparticles [42]. While both magnetic field-mediated gene transfer and Tat peptide-assisted delivery offer attractive features, the combination of the two methods has been documented to provide additional advantages in improving targeted transgene expression [42].

4 Nanotubes for DNA Transfection

Various nanomaterials with unique physical and chemical properties have been used in the biomedical field, with applications for biosensors, biochips, diagnostic devices, implantable medical devices (e.g., prostheses), drug delivery systems, and imaging probes [29]. In particular, carbon-based nanomaterials (e.g., single/multiwalled carbon nanotubes and graphene) have attracted attention owing to their unique properties such as high conductivity, transparency, mechanical strength, and good biological characteristics.

Carbon nanotubes (CNTs) are well-known one-dimensional nanomaterials with high aspect ratio, high surface area and specific mechanical, electrical, and chemical properties [10]. Proteins [25], DNA [40] and RNA [26] were immobilized on the sidewalls of functionalized CNTs, and the interactions between these biomolecule–CNT conjugates with the living cell system were studied. The results showed that CNTs possessed the abilities to penetrate cellular membrane and mediate the internalization of biomolecules into various mammalian cells [11], suggesting ideal application potentials as gene or protein delivery scaffolds.

Was suggested that nanotube uptake occurs via insertion and diffusion through the lipid bilayer of cell membrane. While the uptake mechanism is unclear, it has been consistently reported that well-processed water-soluble nanotubes exhibit no apparent acute cytotoxicity to all living cell lines investigated thus far [37].

Covalent and non-covalent sidewall functionalization of single-walled carbon nanotubes (SWNT) has been actively pursued in recent years aimed at several important goals. The first is to impart solubility to nanotubes in various solvents needed for dispersion, manipulation, sorting and separation. The second is to impart chemical functionality to nanotubes by attaching organic, inorganic or biological species to facilitate the interfacing of nanotubes with other materials for useful composites or bioconjugates.

Functionalization of SWNTs with highly hydrophilic groups has been sought after in order to render nanotubes soluble in aqueous solutions. This would allow interfacing nanotubes with biological systems, potentially leading to an understanding of the biocompatibility of nanotubes and the development of interesting biological applications. It has now been surprisingly found that non-covalent functionalization of SWNTs can be accomplished by binding proteins to the nanotubes by various mechanisms, including strong adsorption of phospholipids grafted with polyethylene glycol (PEG) chains, which renders the nanotubes highly water-soluble.

In addition, several studies have demonstrated the use of complexes of DNA-nanopolymer-carbon nanotubes. [37] demonstrated the improved GFP gene transfection mediated by polyamidoamine dendrimer-functionalized multi-walled carbon nanotubes with high biocompatibility. One drawback associated with such complexes is that the dissociation of the complex is either difficult and/or has not been soundly assessed.

Biocompatibility is an important issue in the development of new nanomaterials for bioapplications. It is of great importance to evaluate cytotoxicities of nanomaterials and to document the information as a database for references. In addition, a fundamental understanding of the interaction of biosystems with nanomaterials at molecular levels, including the induced cellular responses and their effects on the biosystem, is needed [32].

Recently, we have used halloysite clay nanotubes (HCNs) for DNA transfection in sperm cells with high biocompatibility [7]. This experiment is better described in next section.

5 Sperm DNA Nanotransfection: NanoSMGT

Over the past 30 years, several methods to generate transgenic animals have been developed [41]. The most common methods have been pronuclear microinjection, somatic cell nuclear transfer, retroviral vectors, and most recently, embryonic-stem cell transgenesis. The use of sperm for transgenesis has been studied, and several distinct approaches have been developed; however, the efficiency of such sperm-mediated gene transfer (SMGT) needs to be radically improved [12, 13, 27]. The limited uptake of exogenous DNA and its subsequent degradation by sperm, remain the primary factors underlying the low efficiency of this technique [1, 19, 30]. Several approaches have been utilized in order to improve DNA uptake. Methods such as electroporation, lipofection, DNA/DMSO complexes and restriction-enzyme-mediated integration (REMI) have been used; however the frequency of transgenic offspring remains low [6, 8, 21, 43]. Conversely, the more recent use of nanotechnology in the context of SMGT has new possibilities for the efficient delivery of exogenous DNA into sperm.

Recently we have demonstrated the successful use of a cationic dendrimer for exogenous DNA transfection into sex-sorted bovine sperm [8]. In addition we demonstrated that this dendrimer was able to improve the number of internalized plasmids by sperm cells in comparison to other common transfection methods such as lipofection or incubation leading to a significant improvement in production of transgenic bovine embryos [7]. This improvement was also accomplished with no negative effects on cell viability and embryo development confirming the biocompatibility of nanopolymer. The improvement on transgene transmission by this nanostructure could be explained by the fact that exogenous DNA might be protected to DNase digestion.

Functional nanometer-scale containers have been successfully used to deliver pharmaceuticals and nucleic acid molecules into animal cells and tissues. In this regard, halloysite clay nanotubes (HCNs) are of particular interest. They are a natural nanomaterial, with deposits in several countries, including the USA and Brazil [33]. Halloysite clay nanotubes (HCNs) are biocompatible and spontaneously incorporated by cells. The diameter of their inner lumen is compatible with many macromolecules and proteins, allowing the entrapping and subsequent slow release of a range of active agents [44]. These characteristics make this nanomaterial a potentially functional transfection agent for use in bovine SMGT.

As previously described for nanopolymers, HCNs were successful used for plasmid transfection in bovine sperm cells improving the number of transfected plasmids to sperm cells and the number of transmitted plasmids to embryos [7]. These results can be explained by the fact that exogenous DNA might be protected to DNase digestion (see Fig. 1).

Previous studies have shown that nanotubes have the ability to protect bound DNA cargoes (and other cargoes including DNA binding proteins) from enzymatic cleavage, both during and after delivery into cells. Although the mechanisms of protection are not yet well understood [45], it has been hypothesized that silica nanotubes can act as a physical shield that protects the loaded materials from damage, since cellular nucleases/proteins cannot physically access the DNA [9]. Studies using MCF-7 and HeLa cells demonstrated that HCNs are spontaneously captured by cells and become concentrated in the nuclear region [44]; therefore, nanotubes can be internalized by cells and can deliver exogenous DNA directly to the nucleus.

Nanoparticle-mediated gene delivery has recently emerged as a promising tool for gene therapy strategies. Due to their size, they can penetrate practically any tissue, including the blood brain barrier, and tissues protected by tight junctions. For this potential to be accomplished, an ability to target the nanoparticles to

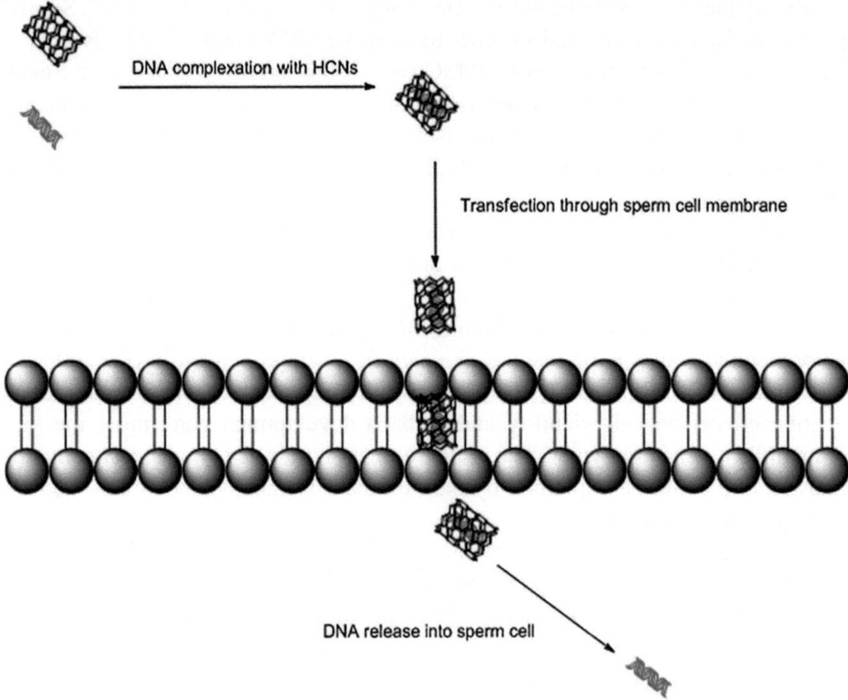

Fig. 1 Scheme of DNA transfection for sperm cells using halloysite clay nanotubes (HCNs). First DNA is complexed with HCNs and then DNA is incorporated and maintained inside of nanotubes being physically protected from DNase digestion. After transfection DNA is released inside of sperm cell and this cells can be used for in vitro fertilization to generate transgenic animals by NanoSMGT

specific sites through surface chemistry would be important [15]. High specificity can be introduced by using biological moieties that process lock-and key interactions, including those observed in antibody-antigen and enzyme-substrate recognition. In surface modification with biomolecules the surface character is critically important, as is the ability for the immobilized molecules to retain their native conformation and binding profile [38]. Owing to their magnetic properties, biocompatibility, and physical properties such as size, shape, and surface characteristics, iron oxide nanoparticles have been recognized as a promising tool for the site-specific delivery of drugs and other bioactive molecules including DNA vectors.

In this sense, magnetic nanoparticles are being increasingly used in a number of biological and medical applications, including cell sorting and transfection [24]. Magnetofection is a new method for gene transfer that involves the use of magnetic force and plasmid DNA (pDNA)/magnetic bead complexes, and it has been developed for enhancing delivery of gene vectors to target cells [20]. For magnetofection, pDNA was interacted with magnetic beads and attracted to target cells by magnetic force in order to accumulate on the cell surface. Generally, the magnetic nanoparticles are made of iron oxide, which is fully biodegradable, coated with specific cationic proprietary molecules varying upon applications. Their association with the gene vectors is achieved by salt-induced colloidal aggregation and electrostatic interaction. The resulting molecular complexes are then targeted to and endocytosed by cells, supported by an appropriate magnetic field. The association of magnetic nanoparticles and several other compounds to improve transfection efficiency has been reported in several studies [4, 23, 42].

Magnetofection for gene delivery in sperm cells was recently demonstrated by [28], producing transgenic boar embryos expressing enhanced green fluorescent protein. The complex formed between plasmid DNA and MNPs was bound on ejaculated boar spermatozoa at a higher efficiency compared to methods using DNA alone or lipofection. They also demonstrated that, when liposomes were used, DNase I greatly reduced the rate of exogenous DNA binding to sperm, whereas when magnetic nanoparticles were used, the reduction in binding was significantly lower. In addition, magnetic nanoparticles bound to exogenous DNA localized either within the plasma membrane or on the plasma membrane, whereas most liposome-bound exogenous DNA localized on the plasma membrane. This study demonstrates the great potential of magnetic nanoparticles for NanoSMGT and consequently for the improvement in the production of transgenic animals or to their use in gene therapy.

6 Conclusions

Clearly the nanostructures could improve the transfection procedures through nanopolymers, nanoparticles and nanotubes that can be used for gene delivery in a large variety of cells, including germ cells and spermatozoa. In the case of spermatozoa, nanotransfectants could be used in a novel technique called NanoSMGT which is a method used to produce transgenic animals or can be used in human gene therapy.

References

1. Anzar, M., Buhr, M.M.: Spontaneous uptake of exogenous DNA by bull spermatozoa. Theriogenology **65**, 683–690 (2006). doi:10.1016/j.theriogenology.2005.06.009
2. Aoki, T., Miyauchi, K., Urano, E., Ichikawa, R., Komano, J.: Protein transduction by pseudotyped lentivirus-like nanoparticles. Gene Ther. **18**, 936–941 (2011). doi:10.1038/gt.2011.38
3. Baldi, L., Hacker, D.L., Meerschman, C., Wurm, F.M.: Large-scale transfection of Mammalian cells. Methods Mol. Biol. **801**, 13–26 (2012). doi:10.1007/978-1-61779-352-3_2
4. Berry, C.C.: Intracellular delivery of nanoparticles via the HIV-1 tat peptide. Nanomedicine (Lond) **3**, 357–365 (2008). doi:10.2217/17435889.3.3.357

5. Bilensoy, E.: Cationic nanoparticles for cancer therapy. Expert Opin Drug Deliv 7, 795–809 (2010). doi:10(1517/17425247).2010.485983

6. Campos, V.F., Amaral, M.G., Seixas, F.K., Pouey, J.L., Selau, L.P., Dellagostin, O.A., Deschamps, J.C., Collares, T.: Exogenous DNA uptake by South American catfish (Rhamdia quelen) spermatozoa after seminal plasma removal. Anim. Reprod. Sci. 126, 136–141 (2011). doi:10.1016/j.anireprosci.2011.05.004

7. Campos, V.F., de Leon, P.M., Komninou, E.R., Dellagostin, O.A., Deschamps, J.C., Seixas, F.K., Collares, T.: NanoSMGT: Transgene transmission into bovine embryos using halloysite clay nanotubes or nanopolymer to improve transfection efficiency. Theriogenology 76, 1552–1560 (2011). doi:10.1016/j.theriogenology.2011.06.027

8. Campos, VF., Komninou, ER., Urtiaga, G., de Leon, PM., Seixas, FK., Dellagostin, OA., Deschamps, JC., Collares, T.: NanoSMGT: transfection of exogenous DNA on sex-sorted bovine sperm using nanopolymer. Theriogenology 75, 1476–1481. (2011c) 10.1016/j.theriogenology.2011.01.009

9. Chen, C.C., Liu, Y.C., Wu, C.H., Yeh, C.C., Su, M.T., Wu, Y.C.: Preparation of fluorescent silica nanotubes and their application in gene delivery. Adv. Mater. 17, 404 (2005). doi:10.1002/adma.200400966

10. Chen, J., Hamon, M.A., Hu, H., Chen, Y., Rao, A.M., Eklund, P.C., Haddon, R.C.: Solution properties of single-walled carbon nanotubes. Science 282, 95–98 (1998)

11. Cheng, J., Fernando, K.A., Veca, L.M., Sun, Y.P., Lamond, A.I., Lam, Y.W., Cheng, S.H.: Reversible accumulation of PEGylated single-walled carbon nanotubes in the mammalian nucleus. ACS Nano 2, 2085–2094 (2008). doi:10.1021/nn800461u

12. Collares, T., Campos, V.F., Leon, P.M.M., Cavalcanti, P.V., Amaral, M.G., Dellagostin, O.A., Deschamps, J.C., Seixas, F.K.: Transgene transmission in chickens by sperm-mediated gene transfer after seminal plasma removal and exogenous DNA treated with dimethylsulfoxide or N, N-dimethylacetamide. J. Biosci. 36, 613–620 (2011). doi:10.1007/s12038-011-9098-x

13. Collares, T., Campos, V.F., Seixas, F.K., Cavalcanti, P.V., Dellagostin, O.A., Moreira, H.L., Deschamps, J.C.: Transgene transmission in South American catfish (Rhamdia quelen) larvae by sperm-mediated gene transfer. J. Biosci. 35, 39–47 (2010). doi:10.1007/s12038-010-0006-6

14. Dai, Z., Gjetting, T., Mattebjerg, M.A., Wu, C., Andresen, T.L.: Elucidating the interplay between DNA-condensing and free polycations in gene transfection through a mechanistic study of linear and branched PEI. Biomaterials 32, 8626–8634 (2011). doi:10.1016/j.biomaterials.2011.07.044

15. de la Fuente, J.M., Berry, C.C., Riehle, M.O., Curtis, A.S.: Nanoparticle targeting at cells. Langmuir 22, 3286–3293 (2006). doi:10.1021/la053029v

16. Dittrich, M., Heinze, M., Wolk, C., Funari, S.S., Dobner, B., Mohwald, H., Brezesinski, G.: Structure-function relationships of new lipids designed for DNA transfection. Chem. Phys. Chem 12, 2328–2337 (2011). doi:10.1002/cphc.201100065

17. Douglas, S.J., Davis, S.S., Illum, L.: Nanoparticles in drug delivery. Crit. Rev. Ther. Drug Carrier Syst. 3, 233–261 (1987)

18. Fischer, D., Bieber, T., Li, Y., Elsasser, H.P., Kissel, T.: A novel non-viral vector for DNA delivery based on low molecular weight, branched polyethylenimine: effect of molecular weight on transfection efficiency and cytotoxicity. Pharm. Res. 16, 1273–1279 (1999)

19. Garcia-Vazquez, F.A., Ruiz, S., Matas, C., Izquierdo-Rico, M.J., Grullon, L.A., De, O.A., Vieira, L., Aviles-Lopez, K., Gutierrez-Adan, A., Gadea, J.: Production of transgenic piglets using ICSI-sperm-mediated gene transfer in combination with recombinase RecA. Reproduction 140, 259–272 (2010). doi:10.1530/REP-10-0129

20. Gersting, S.W., Schillinger, U., Lausier, J., Nicklaus, P., Rudolph, C., Plank, C., Reinhardt, D., Rosenecker, J.: Gene delivery to respiratory epithelial cells by magnetofection. J Gene Med 6, 913–922 (2004). doi:10.1002/jgm.569

21. Harel-Markowitz, E., Gurevich, M., Shore, L.S., Katz, A., Stram, Y., Shemesh, M.: Use of sperm plasmid DNA lipofection combined with REMI (restriction enzyme-mediated

insertion) for production of transgenic chickens expressing eGFP (enhanced green fluorescent protein) or human follicle-stimulating hormone. Biol. Reprod. **80**, 1046–1052 (2009). doi:10.1095/biolreprod.108.070375

22. Hosseinkhani, H., Azzam, T., Tabata, Y., Domb, A.J.: Dextran-spermine polycation: an efficient nonviral vector for in vitro and in vivo gene transfection. Gene Ther. **11**, 194–203 (2004). doi:10.1038/sj.gt.3302159
23. Ino, K., Kawasumi, T., Ito, A., Honda, H.: Plasmid DNA transfection using magnetite cationic liposomes for construction of multilayered gene-engineered cell sheet. Biotechnol. Bioeng. **100**, 168–176 (2008). doi:10.1002/bit.21738
24. Kadota, S., Kanayama, T., Miyajima, N., Takeuchi, K., Nagata, K.: Enhancing of measles virus infection by magnetofection. J. Virol. Methods **128**, 61–66 (2005). doi:10.1016/j.jviromet.2005.04.003
25. Kam, N.W., Dai, H.: Carbon nanotubes as intracellular protein transporters: generality and biological functionality. J. Am. Chem. Soc. **127**, 6021–6026 (2005). doi:10.1021/ja050062v
26. Kam, N.W., Liu, Z., Dai, H.: Functionalization of carbon nanotubes via cleavable disulfide bonds for efficient intracellular delivery of siRNA and potent gene silencing. J. Am. Chem. Soc. **127**, 12492–12493 (2005). doi:10.1021/ja053962k
27. Kang, J.H., Hakimov, H., Ruiz, A., Friendship, R.M., Buhr, M., Golovan, S.P.: The negative effects of exogenous DNA binding on porcine spermatozoa are caused by removal of seminal fluid. Theriogenology **70**, 1288–1296 (2008). doi:10.1016/j.theriogenology.2008.06.011
28. Kim, T.S., Lee, S.H., Gang, G.T., Lee, Y.S., Kim, S.U., Koo, D.B., Shin, M.Y., Park, C.K., Lee, D.S.: Exogenous DNA uptake of boar spermatozoa by a magnetic nanoparticle vector system. Reprod. Domest. Anim. **45**, e201–e206 (2010). doi:10.1111/j.1439-0531.2009.01516.x
29. Kong, D., Cui, Y.: Opportunities in chemistry and materials science for topological insulators and their nanostructures. Nat Chem **3**, 845–849 (2011). doi:1038/nchem.1171
30. Lanes, C.F., Sampaio, L.A., Marins, L.F.: Evaluation of DNase activity in seminal plasma and uptake of exogenous DNA by spermatozoa of the Brazilian flounder Paralichthys orbignyanus. Theriogenology **71**, 525–533 (2009). doi:10.1016/j.theriogenology.2008.08.019
31. Lappalainen, K., Jaaskelainen, I., Syrjanen, K., Urtti, A., Syrjanen, S.: Comparison of cell proliferation and toxicity assays using two cationic liposomes. Pharm. Res. **11**, 1127–1131 (1994)
32. Lay, C.L., Liu, J., Liu, Y.: Functionalized carbon nanotubes for anticancer drug delivery. Expert Rev. Med. Devices **8**, 561–566 (2011). doi:10.1586/erd.11.34
33. Levis, S.R., Deasy, P.B.: Characterisation of halloysite for use as a microtubular drug delivery system. Int. J. Pharm. **243**, 125–134 (2002). doi:S0378517302002740
34. Li, C., Guo, T., Zhou, D., Hu, Y., Zhou, H., Wang, S., Chen, J., Zhang, Z.: A novel glutathione modified chitosan conjugate for efficient gene delivery. J Control Release **154**, 177–188 (2011). doi:10.1016/j.jconrel.2011.06.007
35. Liu, Z., Zheng, M., Meng, F., Zhong, Z.: Non-viral gene transfection in vitro using endosomal pH-sensitive reversibly hydrophobilized polyethylenimine. Biomaterials **32**, 9109–9119 (2011). doi:10.1016/j.biomaterials.2011.08.017
36. Makhluf, S.B., Abu-Mukh, R., Rubinstein, S., Breitbart, H., Gedanken, A.: Modified PVA-Fe3O4 nanoparticles as protein carriers into sperm cells. Small **4**, 1453–1458 (2008). doi:10.1002/smll.200701308
37. Qin, W., Yang, K., Tang, H., Tan, L., Xie, Q., Ma, M., Zhang, Y., Yao, S.: Improved GFP gene transfection mediated by polyamidoamine dendrimer-functionalized multi-walled carbon nanotubes with high biocompatibility. Colloids Surf. B Biointerfaces **84**, 206–213 (2011). doi:10.1016/j.colsurfb.2011.01.001
38. Sajja, H.K., East, M.P., Mao, H., Wang, Y.A., Nie, S., Yang, L.: Development of multifunctional nanoparticles for targeted drug delivery and noninvasive imaging of therapeutic effect. Curr. Drug Discov. Technol. **6**, 43–51 (2009)
39. Shimamura, M., Morishita, R.: Naked plasmid DNA for gene therapy. Curr. Gene Ther. **11**, 433, (2011) BSP/CGT/E-Pub/00089

40. Singh, R., Pantarotto, D., McCarthy, D., Chaloin, O., Hoebeke, J., Partidos, C.D., Briand, J.P., Prato, M., Bianco, A., Kostarelos, K.: Binding and condensation of plasmid DNA onto functionalized carbon nanotubes: toward the construction of nanotube-based gene delivery vectors. J. Am. Chem. Soc. **127**, 4388–4396 (2005). doi:10.1021/ja0441561
41. Smith, K., Spadafora, C.: Sperm-mediated gene transfer: applications and implications. BioEssays **27**, 551–562 (2005). doi:10.1002/bies.20211
42. Song, H.P., Yang, J.Y., Lo, S.L., Wang, Y., Fan, W.M., Tang, X.S., Xue, J.M., Wang, S.: Gene transfer using self-assembled ternary complexes of cationic magnetic nanoparticles, plasmid DNA and cell-penetrating Tat peptide. Biomaterials **31**, 769–778 (2010). doi:10.1016/j.biomaterials.2009.09.085
43. Spadafora, C.: Sperm-mediated gene transfer: mechanisms and implications. Soc. Reprod. Fertil. Suppl. **65**, 459–467 (2007)
44. Vergaro, V., Abdullayev, E., Lvov, Y.M., Zeitoun, A., Cingolani, R., Rinaldi, R., Leporatti, S.: Cytocompatibility and uptake of halloysite clay nanotubes. Biomacromolecules **11**, 820–826 (2010). doi:10.1021/bm9014446
45. Wu, Y., Phillips, J.A., Liu, H., Yang, R., Tan, W.: Carbon nanotubes protect DNA strands during cellular delivery. ACS Nano **2**, 2023–2028 (2008). doi:10.1021/nn800325a
46. Yao, H., Ng, S.S., Tucker, W.O., Tsang, Y.K., Man, K., Wang, X.M., Chow, B.K., Kung, H.F., Tang, G.P., Lin, M.C.: The gene transfection efficiency of a folate-PEI600-cyclodextrin nanopolymer. Biomaterials **30**, 5793–5803 (2009). doi:10.1016/j.biomaterials.2009.06.051
47. Zecchin, D., Di, N.F.: Transfection and DNA-mediated gene transfer. Methods Mol. Biol. **731**, 435–450 (2011). doi:10.1007/978-1-61779-080-5_35

Applications of Carbon Nanotubes in Oncology

Virginia Campello Yurgel, Vinicius Farias Campos, Tiago Collares and Fabiana Seixas

Abstract Nanooncology is based on the use of nanoscale materials to provide tools for cancer detection, prevention, diagnosis and treatment. Due to their unique physical and chemical properties, carbon nanotubes (CNTs) are among newly developed products and are currently of much interest, with a large amount of research dedicated to their novel applications. In cancer research, many advantages of CNTs in drug delivery systems, cellular Imaging, and Cancer Photothermal therapy draw attention. Their physicochemical features enable introduction of several pharmaceutically relevant entities and allow for rational design of novel candidate nanoscale constructs. Thus, a detailed understanding of recent progress in nanooncology, focusing on biomedical research exploring possible application of carbon nanotubes, is required to consider the medical applications of these materials.

1 Introduction

Nanobiotechnology, the application of nanotechnology in life sciences, is starting to show the promise of a great impact on medicine, especially in cancer. Nanooncology is based on the use of nanoscale materials to provide tools for cancer detection, prevention, diagnosis and treatment. Nanoparticles can be engineered to incorporate a wide variety of chemotherapeutic agents and target the delivery of these agents specifically to the tumor site, providing opportunities for

V. C. Yurgel (✉) · V. F. Campos · T. Collares · F. Seixas
Grupo de pesquisa em Oncologia Celular e Molecular, Programa de Pós-Graduação em Biotecnologia, Centro de Desenvolvimento Tecnológico, Universidade Federal de Pelotas, Pelotas, RS, Brazil
e-mail: virginia.yurgel@gmail.com

C. Avellaneda (ed.), *NanoCarbon 2011*, Carbon Nanostructures,
DOI: 10.1007/978-3-642-31960-0_5, © Springer-Verlag Berlin Heidelberg 2013

designing properties that are not possible with other types of drugs. Besides showing potential as a new generation of cancer therapeutics, many types of nanoparticle-based technologies are in development for improve diagnostic imaging of a variety of cancer types. Different kinds of nanocarriers are being investigated for medical applications, such as polymeric nanoparticles, lipid-based carriers, dendrimers, inorganic nanoparticles and carbon nanotubes (CNT) [28].

Nanoparticles are suitable for two tasks required for targeted drug delivery to pathological sites in the body: Passive and active targeting. Passive targeting refers to the accumulation of a drug carrier system at a desired site owing to physico-chemical or pharmacological factors. It benefits from the size of nanoparticles, longevity of the carrier in blood, and the unique properties of tumor vasculature. This approach makes use of the anatomical and functional differences between normal and tumor vasculature, as angiogenic blood vessels in tumor tissues have gaps between adjacent endothelial cells. This characteristic coupled with poor lymphatic drainage induces the enhanced permeability and retention (EPR) effect, which enables macromolecules, including nanoparticles, to pass through these gaps into extravascular spaces and accumulate inside tumor tissues. Another contributor to passive targeting is the microenvironment surrounding tumor cells, since hyperproliferative cancer cells have a high metabolic rate, and the supply of oxygen and nutrients is usually not sufficient to maintain this growth, the use of glycolysis to obtain extra energy results in an acidic environment. Active targeting involves the attachment of a homing moiety, such as a monoclonal antibody or a ligand, to deliver a drug to pathological sites or to cross biological barriers based on molecular recognition processes [51].

Due to their unique physical and chemical properties, carbon nanotubes are among newly developed products and are currently of much interest, with a large amount of research dedicated to their novel applications. Thus, a detailed understanding of recent progress in nanooncology, focusing on biomedical research exploring possible application of carbon nanotubes as drug delivery carriers and diagnostic devices, is required to consider the medical applications of these materials.

2 Cancer Therapy

Cancer arises through a multistep mutagenic process, allowing cancer cells to acquire properties of unlimited proliferation potential and self-sufficiency in growth signals and resistance to both anti-proliferative and apoptotic cues. The current treatments are surgery, radiation therapy and chemotherapy, all of which cause damage to healthy cells [45].

The efficacy of chemotherapy is often limited by resistance mechanisms, including decreased uptake into cancer cells, enhanced detoxification and efficient elimination of the drugs from cells. Moreover, severe side-effects can restrict the drug dose or even cause termination of the therapy. If not the discovery of new drugs, the strategy to increase the anticancer activity and minimize resistance should be improving delivery of conventional chemotherapeutics [3].

Chemotherapeutics have limitations due to the lack of selectivity and severe toxicity, decreasing the anticancer dose is highly favorable for lowering toxic side effects of drug therapy to the normal organs and tissues. In this sense, Nanotechnology, through nanoscaled carriers for drug transport with the goal to improve the drug efficacy by lower doses and fewer side-effects, is a promising instrument [3, 71]. In addition, limited solubility, poor distribution among cells, inability of drugs to cross cellular barriers, and especially a lack of clinical procedures for overcoming multidrug resistant (MDR) cancer, all limit the clinical administration of chemotherapeutic agents [29].

The application of nanomaterials as drug carriers can improve anticancer therapeutic efficacy by both passive and active targeting mechanisms [51]. Passive target is induced by tumor-specific EPR-effect, which warrants the development of nanomedicine and is applicable for any biocompatible macromolecular compounds above 40 kDa, even larger than 800 kDa. The drug concentration in tumor compared to that of the blood can be usually as high as 10–30 times, and it is not just passive targeting for momentary tumor delivery, but it means prolonged drug retention for more than several weeks or longer [46].

Carbon nanotubes (CNTs) have been proposed as multipurpose innovative carriers for drug delivery and diagnostic applications. Their physicochemical features enable introduction of several pharmaceutically relevant entities and allow for rational design of novel candidate nanoscale constructs for drug development [63]. CNTs loaded with drugs (which can be regarded as macromolecular agents) can extravasate in tumor tissues over time; the concentration in tumor will reach several folds higher than that of the plasma [93].

Thus, Carbon nanotubes can help greatly in treating the cancer cells by delivering chemotherapeutic agents achieving better uptake by malignant cells without affecting collateral tissues. Consequently, nanotubes potentially lower the dose of drug by localizing its distribution at the tumour site [5].

3 Carbon Nanotubes

Carbon nanotubes (CNTs) are hollow graphitic nanomaterials with very high aspect ratios. Can be described as rolled sheets of graphene with lengths from several hundred nanometers to several micrometers, which can be single-walled (SWNTs) with diameters of 0.4–2 nm, or multi-walled (MWNTs) consisting of 2–30 concentric tubes positioned within one another, with outer diameters ranging from 2 to 100 nm [14, 37] (Fig. 1). Several reviews have been published on synthesis and structural conformation of CNTs [5, 72].

The three main techniques to produce CNTs are electric arc discharge, laser ablation and chemical vapor deposition. These methods involve synthesis at high temperature, pressure and the use of reaction catalysts, leading to fine structures of

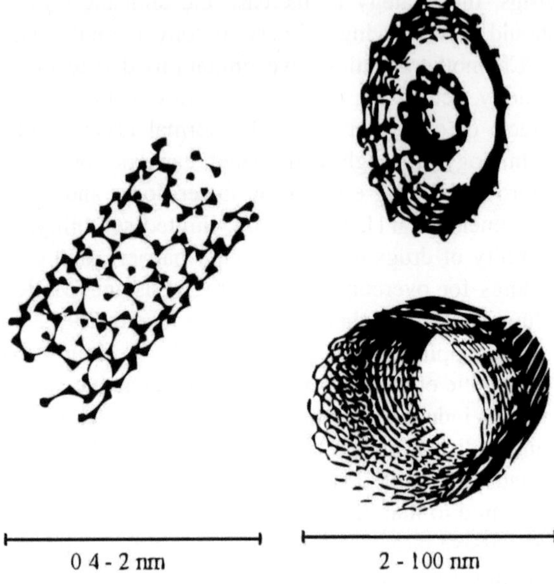

Fig. 1 Single-walled nanotubes and multi-walled nanotubes showing typical diameters

CNTs along with some synthesis induced impurities like graphitic debris and catalytic particles, and there are various important methods for purification [61].

Recently, efforts have been devoted to exploring the potential biological applications of CNTs motivated by their interesting properties [14]. Noticeably, biomedical research is exploring possible application of carbon nanotubes as drug delivery carriers and diagnostic devices [37]. These materials can be used for hyperthermic ablation of cancer cells due of their strong optical absorption in the NIR wavelength region, as well as for drug delivery to cancer cells owing to their high surface areas [26]. The advantages of carbon nanotubes in biomedical applications rely on the fact that they can be easily internalized into a wide variety of cell types and through several mechanisms [34], being able to act as delivery vehicles for a variety of molecules relevant to therapy and diagnosis.

Since cellular internalization of CNTs takes an important role in targeting therapy for cancer, understanding the exact internalization mechanisms is crucial for pharmaceutical application of CNTs [65]. The suggested mechanisms by which CNTs cross the cell membrane are still being debated, with two major intracellular uptake mechanisms being proposed: endocytosis/phagocytosis and nanopenetration. Whereas it was demonstrated that the mechanism for the cellular uptake of CNTs is endocytosis [30], it was also reported that the CNTs are able to enter the cells by a nanopenetration mechanism [11]. When caring a chemotherapeutic agent, by virtue of endocytosis, the CNTs can be taken up by the cell before the chemotherapeutic drugs are cleaved off CNTs, and thus, targeting delivery is realized.

Regarding cellular dynamics, Zhou et al. [95], using confocal laser scanning microscopy, observed SWNTs dispersed in phospholipid-polyethylene glycol

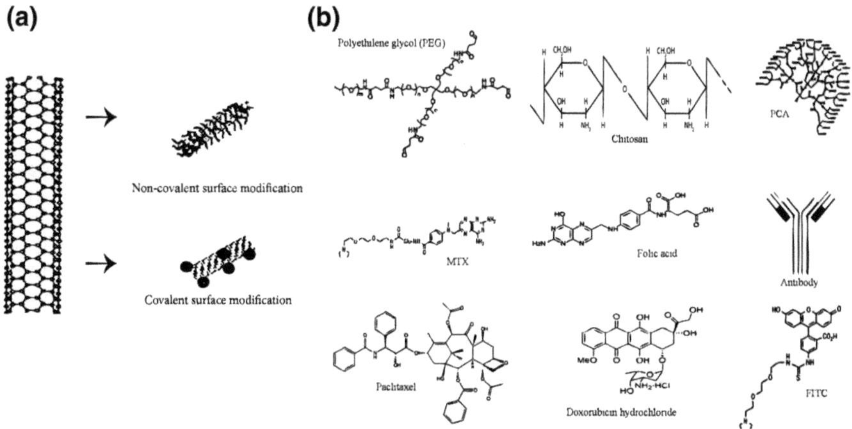

Fig. 2 Functionalization schemes: **a** schematic illustration for covalent and non-covalent modifications of carbon nanotubes; **b** some molecules that can be used for engineering carbon nanotubes formulations

(PLPEG) and conjugated with different molecules, in tumor, normal, and macrophage cells, to determine the subcellular localization and to study the transmembranal mechanism of SWNTs. They found that SWNT-PL-PEG-FITC was localized in the mitochondria of both tumor and normal cells due to mitochondrial transmembrane potential. The mitochondrial SWNT-PL-PEG, when irradiated with a near infrared light, could induce cell apoptosis due to mitochondrial damages. And also, SWNT-PL-PEG could be localized in different subcellular components by conjugations of different molecules. Subcellular localization of surface-modified SWNTs depends on how they enter the cells. When the conjugated molecules can specifically target tumor cells, they are bound to the cell surface; their pathway into the cell is through endocytosis, resulting in the lysosomal distribution. While for non cell-targeting molecules the entry mode depends on the properties of the conjugated molecules [95].

Carbon nanotubes are insoluble in most organic or aqueous solvents, are hardly dispersed in aqueous solutions and have a strong tendency to interact hydrophobically and aggregate [33]. Therefore, for biological application the nanotube surface needs to be modified. Hence, for biomedical and pharmaceutical applications, various chemical modifications, by different strategies, have been used to solubilize and disperse carbon nanotubes in water, and make then biologically compatible. Such modifications can be performed by covalent attachment of chemical groups, supramolecular adsorption or wrapping of functional molecules, or yet the filling of the inner cavity of the nanotube [72]. Furthermore, the edges of the tube holes have oxidized functional groups where covalent attachment of chemicals is possible [2].

The modifications can improve biocompatibility and capability to penetrate cell membranes, enhance effective transport of molecules into the cytoplasm via cell membrane without producing a toxic effect, and also offer a flexible platform for further derivation processes [7]. Thereby, modification of carbon nanotubes through covalent or non-covalent functionalization of their external walls is a key step for biomedical applications because a wide variety of active molecules can be linked to a functionalized carbon nanotube [60] (Fig. 2).

The two main methodologies are based on the non-covalent coating of nanotubes with amphiphilic molecules (like polymers or surfactants), and the covalent functionalization of the nanotube surface by grafting various chemical groups directly onto the backbone [33, 72, 73] Covalent functionalizations are strong and offer the possibility of introducing multiple functionalities. By this type of functionalizations, defective carbon atoms on the sidewall or at the end of CNTs can be oxidized by strong oxidants to generate carboxylic acid groups or carboxylated fractions, which can be chemically modified via amidation or esterification, and then various polymers, metals, and biological molecules can be grafted to the surface [93].

The method of introducing carboxylic and other oxygen-containing groups to the inert carbon nanotube allows the covalent attachment of functional molecules, thus offering new opportunities for applications in science [21], but it also generates so-called "oxidation debris" by breaking up CNTs during oxidation. Heister et al. [23] demonstrated that the same acid oxidation protocol leaves some CNT samples nearly unaltered, whereas others are significantly broken up, leading to altered CNT dimensions and surface properties, which influence their dispersion in salt solutions, cellular growth media, and human plasma. A similar diversity was found for the effect of the removal of oxidation debris, which leaves the tubes with a clean surface, but also impairs dispersion stability. However, functionalization of CNTs with appropriate biomolecules allows for tailoring of their surface properties and is shown to improve dispersion [23].

Compared to covalent functionalization of the CNT sidewall, the supramolecular non-covalent functionalization approach has the advantage of establishing strong interactions with the nanotubes without altering their electronic nature and peculiar features, like optical properties useful for various biological imaging and sensing applications [17]. Furthermore, another strategy relies on the fact that CNTs have a high affinity for DNA and RNA, and are able to condense DNA to varying degrees, which can grant aqueous solubility and provide gene transfer vector systems [70].

Clearly, for a good biomedical performance the degree of aggregation and individualization of nanotube materials in biological fluids has an important role. In addition, the blood clearance is one of the parameters that must be determined for the development of any pharmacological agent, and this parameter also depends on the nanotube surface modification [37].

Polyethylene glycol (PEG) has been a traditional mean of functionalization of carbon nanotubes in drug delivery systems [20]. The presence of PEGchains endows CNTs with a hydrophilic coating, which also reduces protein adsorption and phagocytosis. Thus, PEGylation is one of the most effective methods to prolong the blood-circulation time, due to the properties of polyethylene glycol to overcome the phagocytic activity of the reticulo-endothelial system [17, 37].

The long circulation time is generally desired for nanovehicles as it allows nanomaterials to repeatedly pass through tumor vasculatures facilitating passive targeting to the cancer cells by the enhanced permeability and retention effect of tumor blood vessels [47]. Liu et al. [38] systematically studied the relationship between polymer surface coatings and in vivo behaviors of SWNTs. By controlling the PEGylation degree they achieved an optimal blood circulation half-life of SWNTs at 12 e 13 h, which afforded relatively low RES accumulation, high tumor uptake and low skin retention of SWNTs in mice, ideal for in vivo cancer treatment.

Lately, other strategies have been proposed. The functionalization of multi-walled carbon nanotubes (MWNTs) with highly hydrophilic and biocompatible poly(vinyl alcohol) (PVA) increase their aqueous solubility, and shown no obvious toxicity [67]. Also, hyperbranched poly citric acid (PCA), with a high capacity for conjugation to drug molecules, was used for the functionalization of MWNTs instead of polyethylene glycol with limited arms that is commonly used to conjugate to other molecules for drug delivery systems [71].

Poly citric acid is a highly biocompatible hydrophilic polymer [74] that beyond decreasing hydrophobicity of CNTs is also a highly functional polymer with a large number of carboxylic functional groups that confer a high loading capacity [71]. As also reported in a study using polyglycerol where MWNTs assumed a circular form [1], this kind of functionalization causes conformational changes from linear toward curved nanotubes, due to the hydrophobicity of CNTs and high hydrophilicity of poly citric acid in aqueous solutions, consequently reducing their size [71].

Concerning active targeting, researchers have been developing CNTs coupled with cancer cell-specific targeting moieties to enhance CNTs uptake by cancer cells while limiting uptake by normal cells [6, 31]. To drive CNTs for selective internalization into cancer cells, it is advantageous to functionalize them with antibodies of antigens overexpressed on the cancerous cell surface or with specific ligands that recognize receptors on the cancerous cell. For this purpose, some targeting molecules have been reported, such as Folic acid [16], monoclonal antibodies [50], epidermal growth factor [6], Rituxan (to selectively recognize CD20 cell surface receptor on B-cells) and Herceptin (to recognize HER2/neu receptor positive breast cancer cells) [83].

As for other molecules, to conjugate immunoglobulins (Ig) to CNTs, covalent bonding and noncovalent interactions strategies have been reported, and the Immuno-CNT constructs (antibody (Ab)–CNT conjugates) can modulate immunological functions, provide specific targeting, and enhance the efficacy of antitumor therapies [76]. CNTs combined with the antibody anti- P-glycoprotein (which is overexpressed on multidrug resistant cells) and loaded with an anti-neoplastic drug, could not only specifically recognize the multidrug resistant cells, but also demonstrated the effective loading and controllable release performance of the drug [36]. It was also reported CNTs constructs appended with a radiotherapeutic agent and the tumor neovascular-targeting antibody E4G10, thus increasing specificity and limiting the cellular damage of healthy tissues [66].

Another promising application of CNTs in biology and medicine is the development of advanced biosensor devices [75, 79]. Recently, Park et al. [59] synthesized D-(+)-Galactose-conjugated single-walled carbon nanotubes (SWNTs) for use as biosensors to detect the cancer marker galectin-3. High levels of this circulating marker are correlated with an increased potential for malignancy in several types of cancer, and the electrochemical sensitivity measurements of the D-(+)-galactose-conjugated SWNTs differed significantly between the samples with and without galectin-3, indicating that conjugated SWNTs are potentially useful electrochemical biosensors for marker detection. In attempt to diagnose multi drug resistance (MDR) in cancer, which is responsible for a large portion of chemotherapeutic failures, Zhang et al. [91] demonstrated a new strategy using an electrochemical sensor based on carbon nanotubes aiming a fast and sensitive method of assessment of cancer MDR, and thus guiding how to reverse it in clinic therapy.

For gene therapy, the generation of CNT- DNA complexes can be especially beneficial because the presence of DNA in this complex stabilizes the carbon nanoparticle in aqueous suspensions and the DNA molecule is protected from unwanted enzymatic cleavage and nucleic acid binding protein interference when attached to the carbon nanotubes [70, 88]. A gene delivery concept based on ethylenediamine-functionalized single-walled carbon nanotubes (f-SWNTs) using the oncogene suppressor p53 gene as a model gene was successfully tested in vitro in MCF-7 breast cancer cells [32]. They demonstrated the ability of f-SWNT-p53 complexes to act as gene delivery vehicles and facilitate the delivery of p53 plasmid DNA, resulting in the expression of this protein. This system could be the foundation for novel gene delivery platforms based on the unique structural and morphological properties of multi-functional nanomaterials and be a tool for the preliminary screening of different genes for their ability to affect cancer cell growth [32].

There is evidence that carbon nanotubes can produce immune responses when covalently linked to highly immunogenic peptide sequences [58]. Moreover, carbon nanotube-peptide constructs can improve the immunogenicity of a weakly immunogenic clinically relevant cancer-associated peptide, and therefore they are promising tools to explore ways to improve vaccine therapy against cancers [77].

In conclusion, as other nanomaterials, the biocompatibility and applicability of CNTs depends on controllable properties such as size and morphology, and also additives and surface modifications, since interactions between carbon nanotubes and living cells are strongly dependent on surface chemistry. In cancer research many advantages of CNTs in drug delivery systems, cellular Imaging and Cancer Photothermal therapy, draw attention.

4 Drug Delivery

Often, chemotherapeutic agents offer a limitation of solubility and cell-penetration ability. Plus, the systemic toxicity caused by lack of selective limits the clinical applications. Therefore, development of an effective drug delivery system becomes an active area of research. The bioavailability of a drug can be maximized by the attachment to a suitable carrier. The therapeutic efficacy can be improved and side effects reduced because of facilitated transportation of drugs to the desired target.

As mentioned, CNTs are potential drug delivery vectors due to their ability to cross cell membranes easily and their high aspect ratio as well as high surface area, which provides multiple attachment sites for drug targeting. Also, they are stable, inert and present higher surface area-to-volume ratio than spheres [43], which provides higher loading capacity for guest molecules, suggesting the potential utility of these materials as carriers in drug delivery systems that require higher loadings of therapeutic agents [44, 49].

Carbon nanotubes offer many potential advantages over other types of nanoparticles for cancer therapy. Their unique physical properties permit efficient electromagnetic stimulation and highly sensitive detection using various imaging modalities. Their large surface area and internal volume also allows drugs and a variety of small molecules, such as contrast agents, to be loaded onto the nanotube. Carbon nanotubes have been used to halt tumor growth in the context of various therapeutic modalities including chemotherapy, hyperthermia and gene silencing [33].

SWNTs are promising carriers for drug delivery since they are relatively safe inorganic materials which are capable to penetrate cell membranes, can be covalently functionalized with small molecules, and several anticancer drug molecules have been transported into different types of cells by appropriately functionalized SWNTs [37]. However, MWNTs seem to be more suitable for the encapsulation of drugs because of their wider inner diameter, and the outer shells can be functionalized without destroying the side walls [3].

In order to perform the drug load, different approaches can be followed, ranging from covalent or noncovalent attachment of drug molecules to the sidewalls of functionalized CNTs [40, 87] to the incorporation of drug molecules into their interiors [2, 22]. CNTs have been proved to be versatile drug carriers, increasing the activity of various drugs used in cancer treatment, such as doxorubicin [25], carboplatin [3] and paclitaxel [71].

The chemical conjugation changes the chemistry entities of the drugs, especially when the drugs are covalently conjugated via non-biodegradable linkages [35], so, physically loading the drugs onto CNTs could be preferred because of no changes in the chemical entities, but this approach may be limited to certain drugs [22, 43]. Lay et al. [35] set up an approach to physically load Paclitaxel (PTX) onto the side walls of CNTs by immersing PEG-g-SWNTs and PEG-g-MWNTs in a saturated solution of PTX in methanol. PTX loaded PEG-g-CNTs could be well dispersed in aqueous solution without aggregation. Moreover, PTX could be sustained released from PEG-g-CNTs faster than free PTX, with a high in vitro efficacy to kill cancer cells.

A successfully synthesized MWNT–PVA could form stable complexes with the Camptothecin (CPT) via noncovalent interactions and exhibited higher cytotoxic activity compared to free CPT alone [67]. Therefore, functionalized MWNTs can be load and mediate delivery of poorly water-soluble anticancer drugs, and as water soluble complex can significantly improve the activities of the drug, enhancing the cellular uptake [67]. Another strategy is functionalization with the safe and nontoxic polysaccharide Chitosan. When the Chitosan is functionalized on the surface of CNTs, the cells become attached to the sidewalls of the nanotubes, resulting in the desired targeted release to the cells, with improved drug absorption [5].

Aiming to minimize the disadvantage of systemic administration of the chemotherapeutic drug Doxorubicin (DOX), which kills healthy cells, it was synthesized a nanocarrier by non-covalent attachment of DOX to CNT surface followed by encapsulation of CNTs with folic acid-conjugated Chitosan, resulting in a system with characteristics of both controlled release and specific targeted, given that folate receptors are overexpressed on cancer cells [25]. There was an initial burst release and then a gradual release of DOX, which is a desirable characteristic for therapy. Also, the system had a higher drug release at acidic medium, which promotes higher drug release caused by partial dissociation of hydrogen bonding interaction between DOX and SWCNT. The superior controlled release from chitosan-folic acid encapsulated nanocarrier in contrast to non-encapsulated can be attributed to degradation of Chitosan and diffusion through the Chitosan shell, and also folic acid–DOX hydrogen bonding [25].

Nanoparticles can also be explored as nanocontainer-based drug depots, which constantly provide the active drug, being very useful in case of unstable drugs that become inactivated over time. CNTs loaded with the alkylating chemotherapeutic drug Carboplatin (CP), which is a second- generation platinum agent, showed stronger activity than free Carboplatin [3]. The capability of carboplatin to induce apoptosis was doubled when the drug was transported by CNT–CP, while no significant toxicity was found after cell exposure to unloaded CNTs. Furthermore, CNT–CP continuously released their payload during the observation period, showing a maximum release of platinum of 68 % at day 14 without reaching a plateau [3].

Recently, Sobhani et al. [71] proposed a novel drug delivery system for cancer chemotherapy based on multiwalled carbon nanotubes functionalized with hyperbranched poly citric acid (MWNT-g-PCA) and covalently attached to the commonly used potent chemotherapy drug Paclitaxel (PTX), to produce a MWNT-g-PCA-PTX conjugate. This complex is taken up by cells through endocytosis, where the cleavable ester bond between Paclitaxel and poly citric acid is hydrolyzed, and Paclitaxel is released into the cytoplasm, showing a higher cytotoxic effect than free Paclitaxel. That could be related to the increased cell penetration of Paclitaxel when conjugated. Moreover, the MWNT-g-PCA alone had no significant effect on cell viability, suggesting that the cytotoxicity was caused by the conjugated Paclitaxel only. The improved cytotoxic effect on the cancer cell lines

Table 1 Overview of recent research on carbon nanotubes in drug delivery systems

Type of CNT	Functionalization	Drug	Biological system	References
SWNT	Phospholipid-branched polyethylene glycol	Paclitaxel	in vivo	[40]
SWNT	Epidermal growth factor	Cisplatin	in vitro/in vivo	[6]
MWNT	Hydrophilic diaminotriethylene glycol	Hydroxycamptothecin	in vitro/in vivo	[87]
SWNT	Polysaccharide materials and Folic acid	Doxorubicin	in vitro	[92]
SWNT/ MWNT	Polyethylene glycol	Paclitaxel	in vitro	[35]
SWNT	Antibody of P-glycoprotein	Doxorubicin	in vitro	[36]
MWNT	Ethylene glycol-b-propylene sulfide	Doxorubicin	in vitro	[17]
SWNT	Folate-chitosan conjugate	Doxorubicin hydrochloride	in vitro	[25]
MWNT	Poly(vinyl alcohol)	Camptothecin	in vitro	[67]
MWNT	Hyperbranched poly citric acid	Paclitaxel	in vitro	[71]
MWNT	CoFe2O4	Doxorubicin	in vitro	[86]
MWNT	Poly(acrylic acid) and Fe3O4	Gemcitabine	in vitro/in vivo	[90]

may be due to higher cell penetration of the conjugated nanotubes, a behavior that could be associated to amphiphilicity of the MWNT-g-PCA for improved cell wall interaction compared with absolute hydrophobic Paclitaxel, also to the unique conformation of the carbon nanotube-based nanocarrier, which can penetrate into various cells. Furthermore, the release of the drug from the conjugate was higher at an acid pH, which is suitable for the release of drugs in tumor sites [71].

Intending to fight against metastases, Yang et al. [90] used magnetic carbon nanotubes mMWNT as chemotherapeutic agent vehicles to targeted cancer metastatic lymph nodes under the guide of implanted magnet, and showed successful application of intra-lymphatic delivery of Gemcitabine (GEM) using mMWNTs. Di Crescenzo et al. [17] proposed the use the biocompatible amphiphilic diblock copolymer poly (ethylene glycol-b-propylene sulfide) ($PEG_{44}PPS_{20}$), in order to assess the ability of MWNTs dispersed with $PEG_{44}PPS_{20}$ to assist and direct the entry of the desired amount of Doxorubicin (DOX) in cancer cells and enhance its cytotoxic activity. The authors demonstrated that $PEG_{44}PPS_{20}$ was capable of efficiently and stably dispersing CNTs and of finely tuning the DOX loading onto the nanotube surface. The $PEG_{44}PPS_{20}$ coated MWNT/DOX complex exhibit efficient cell internalization and enhanced cytotoxic activity compared to both DOX alone and DOX-loaded copolymer micelles.

Zhang et al. [92] described a system employing two different polysaccharides, sodium alginate (ALG) and chitosan (CHI), and further functionalization with targeting group folic acid (FA) and an anticancer drug DOX. The complete system displayed excellent stability under physiological conditions, but at reduced pH typical of the tumor environment, intracellular lysosomes and endosomes, the DOX was efficiently released and enters the cell nucleus inducing cell death.

Using the antibody (Ab) target approach, Venturelli et al. [76] designed and prepared new Ab–CNT constructs with anti-MUC1 Ab covalently conjugated to carbon nanotubes. Anti-MUC1 Ab recognizes the mucin 1 (MUC1) receptor which is overexpressed on a series of human cancer cells, and it was chosen as an active targeting molecule in view of the development of a multimodal CNT-based hybrid with therapeutic properties. They obtained conjugates that formed stable homogeneous dispersions under physiological conditions and can be useful for biomedical investigations.

In summary, the transporting capabilities of carbon nanotubes combined with appropriate surface modifications and their unique physicochemical properties can lead to a new kind of nanomaterials for cancer treatment [29]. An overview of recent research on applications of CNTs as drug delivery systems in oncology is shown in Table 1.

5 Cellular Imaging and Cancer Photothermal Therapy

Carbon nanotubes have unique electrical, thermal and spectroscopic properties that offer advances in the detection and monitoring of diseases [33]. Besides having a surface area that allows efficient loading of multiple molecules along the length of the nanotube sidewall [43], the intrinsic optical and electrical properties of CNTs can be utilized for multimodality imaging and therapy.

For therapeutic hyperthermia, heating of an anatomical area without damaging the surrounding tissue is difficult. The potential approach to solving this problem involves the use of nano-sized biocompatible particles that convert absorbed electromagnetic energy to heat after they localize in the tumor [48]. CNTs have the ability to absorb near-infrared (NIR) radiation (700–1100 nm) and then convert it into heat, which provides an opportunity to new strategies for cancer thermal therapy [31]. Also, biological systems are known to be transparent to the same spectral window, so, this can be used for optical imaging of nanotubes inside living cells, with low autofluorescence background of cells and tissues [83].

Nanotubes play an important role in diagnostic procedures by helping in the imaging of organs, by identifying the site of action of drugs in targeted delivery systems, and by having great potential to act as a contrast agent in imaging and identification of cancer cells [62]. Thus, intrinsic fluorescence properties of carbon nanotubes enable their application as biological imaging agents [42], their NIR photoluminescence property can be used for in vitro cell imaging [83], they can be used for photothermal therapy [12, 31], photoacoustic imaging [15], and to deliver therapeutic drugs with externally controlled release capabilities [31].

Magnetic CNTs have shown promising results as a Magnetic resonance imaging (MRI) contrast agent with high nuclear magnetic resonance relativities, little cytotoxicity and high cell-labelling efficiency [4]. It was demonstrated that poly(-acrylic acid) functionalized multi-walled carbon nanotubes (PAA-g-MWNTs) decorated with magnetite nanoparticles (Fe3O4) can be efficiently taken up by lymphatic vessels and delivered to regional lymph nodes in vivo with little toxicities [89].

Carbon nanotubes are capable of photo-excitation. Since they are photoabsorbers, combining laser irradiation can enhance thermal deposition and targeted destruction of cancer cells. Selectively target CNTs, by the use of antibodies, associated with NIR radiation showed success in treating cancer cells both in vitro and in vivo [9, 94]. To destroy cancer cells using hyperthermia, target CNTs uptake only by cancerous cells provide an opportunity to locally heat the carbon nanotubes using infrared laser radiation, resulting in thermal destruction of cancer cells without harming surrounding cells [31].

For laser cancer therapy and improved diagnostic imaging, it has been suggested that a nanomaterial known as a single walled carbon nanohorn (SWNH) offer many advantages. SWNHs are much like SWNTs, but short and conically-shaped sealed structures, with an overall diameter ranging between 50 and 100 nm [54]. One important aspect is that SWNHs are typically produced by laser ablation of pure graphite target samples, eliminating toxicity associated with the presence of metal catalysts [52]. Inclusion of SWNHs may allow the treated tissue to absorb more therapeutic laser light, enhancing the ability to produce lethal temperature elevations necessary for tumor destruction and inhibition of tumor recurrence, selectively increasing the temperature while minimizing thermal impact on locations where NIR laser light is directed without SWNHs [84].

Since highly proliferative tumors have the capacity to create albumin deposits, cancer cells overexpress specific human serum albumin (HSA) receptors and are able to internalize large amounts of albumin, Iancu et al. [27] proposed a method for the noncovalent functionalization of multiwalled carbon nanotubes with HSA for the selective targeting and laser-mediated necrosis of liver cancer cells, showing interesting results. Innovatively, Mocan et al. [53] presented a new method of selective nanophotothermolisys of pancreatic cancer (PC) using MWNTs functionalized with HSA in an original designed model of living PC, using ex vivo-perfused pancreatic specimens that were surgically removed from patients with ductal adenocarcinoma. They obtained a selective photothermal ablation of the malign tissue based on the selective internalization of MWCNTs with HSA inside the pancreatic adenocarcinoma after the ex vivo intra-arterial perfusion.

Through the attachment of an MRI contrast agent onto CNTs in a suitable way, the obtained CNT-based hybrids can be used as an MRI contrast agent and a drug carrier simultaneously. In this sense Wu et al. [86] develop a solvothermal method to synthesize MWCNT/CoFe2O4, and in vitro experiments revealed that the hybrids displayed low cytotoxicity and an excellent MRI enhancement effect on cancer cells. In addition, the hybrids showed a high loading capacity for the anticancer drug DOX, were stable, and allowed fast drug release in acidic environments.

6 Toxicity

Whereas the potential multifunctionality of carbon nanotubes has been shown sufficiently, their toxicity is still controversially discussed. The rapid progress in the development and use of nanomaterials is not yet matched by adequate toxicological investigations. Growth in manufacturing of nanomaterials requires knowledge and control of the possible adverse health effects that may take place throughout the production and use [68]. Thus, the potential toxic effects for the environment and for health have become an issue of concern. The published toxicity data of CNTs are not solid and even conflicting [13, 18, 39, 41, 57, 69, 80, 82, 85].The variation may be due to a wide range of tube diameters, lengths, chiralities, catalyst content, and also different functionalizations, bringing the need of standardization of protocols to establish accurately the toxicity and long term fate of the CNTs.

The diameter of tubular nanomaterials is an important determinant for their loading capacities [24] and an association of the toxicity of multiwalled nanotubes and their diameter was reported [81]. Several studies evaluated toxicology utilizing pristine nanotubes, mostly poor-quality aqueous dispersions, which logically are the most difficult to handle biologically [8]. Moreover, residual metal catalyst may play an important role in the cytotoxicity [19, 78], since oxidative stress is proposed as a key mechanism of CNT-induced toxicity, and it is usually linked to the metallic impurities [64]. Then, designed functionalized CNTs without residual heavy metals, tend to be biocompatible and nontoxic at cellular level [18, 19, 55, 56], offering the potential exploitation of nanotubes for drug administration.

Lately, Burke et al. [10] assessed the potential thrombogenic effects of functionalized MWCNTs in vitro and in vivo, and found that the thrombogenic potential can be substantially moderated through covalent functionalization. The systemic administration of covalently functionalized MWCNTs does not initiate a strongly pro-coagulant state, which is an important factor for their clinical use.

Considering interference with dye-based viability assays, agglomeration issues related to the method of dispersion, and oxidant stress due to metal contamination of CNT, the data available favor the conclusion that well-dispersed, purified CNTs exhibit relatively low cytotoxicity in vitro [68]. So, to consider clinical applications it is important to understand pharmacological and toxicological properties, evaluate risk/benefit ratios, identify which physico-chemical characteristics of carbon nanotubes are capable of driving the toxic responses, and design of carbon nanotubes that are biocompatible and safe.

7 Conclusions

Nanosystems hold great potential to overcome many of the present obstacles of therapy and diagnoses in oncology, and carbon nanotubes have been actively explored as multipurpose innovative versatile carriers because of their great material properties. The uses of CNTs as bioactive molecules are still at an early research stage and face some challenges, but their unique physical and chemical properties hold great hope for cancer management. Overall, the results clearly indicate potential applications of CNTs in oncology, making important steps towards safe medical applications of CNTs.

References

1. Adeli, M., Mirab, N., Zabihi, F.: Nanocapsules based on carbon nanotubes-graft-polyglycerol hybrid materials. Nanotechnology **20**, 485–603 (2009). doi:10.1088/0957-4484/20/48/485603
2. Ajima, K., Murakami, T., Mizoguchi, Y., Tsuchida, K., Ichihashi, T., Iijima, S., Yudasaka, M.: Enhancement of in vivo anticancer effects of cisplatin by incorporation inside single-wall carbon nanohorns. ACS Nano **2**, 2057–2064 (2008). doi:10.1021/nn800395t
3. Arlt, M., Haase, D., Hampel, S., Oswald, S., Bachmatiuk, A., Klingeler, R., Schulze, R., Ritschel, M., Leonhardt, A., Fuessel, S., Buchner, B., Kraemer, K., Wirth, M.P.: Delivery of carboplatin by carbon-based nanocontainers mediates increased cancer cell death. Nanotechnology **21**, 335101 (2010). doi:10.1088/0957-4484/21/33/335101
4. Bai, X., Son, S.J., Zhang, S., Liu, W., Jordan, E.K., Frank, J.A., Venkatesan, T., Lee, S.B.: Synthesis of superparamagnetic nanotubes as MRI contrast agents and for cell labeling. Nanomedicine (Lond) **3**, 163–174 (2008). doi:10.2217/17435889.3.2.163
5. Beg, S., Rizwan, M., Sheikh, A.M., Hasnain, M.S., Anwer, K., Kohli, K.: Advancement in carbon nanotubes: basics, biomedical applications and toxicity. J. Pharm. Pharmacol. **63**, 141–163 (2011). doi:10.1111/j.2042-7158.2010.01167
6. Bhirde, A.A., Patel, V., Gavard, J., Zhang, G., Sousa, A.A., Masedunskas, A., Leapman, R.D., Weigert, R., Gutkind, J.S., Rusling, J.F.: Targeted killing of cancer cells in vivo and in vitro with EGF-directed carbon nanotube-based drug delivery. ACS Nano **3**, 307–316 (2009). doi:10.1021/nn800551s
7. Bianco, A., Kostarelos, K., Prato, M.: Applications of carbon nanotubes in drug delivery. Curr. Opin. Chem. Biol. **9**, 674–679 (2005). doi:10.1016/j.cbpa.2005.10.005
8. Boczkowski, J., Lanone, S.: Potential uses of carbon nanotubes in the medical field: how worried should patients be? Nanomedicine (Lond) **2**, 407–410 (2007). doi:10.2217/17435889.2.4.407
9. Burke, A., Ding, X., Singh, R., Kraft, R.A., Levi-Polyachenko, N., Rylander, M.N., Szot, C., Buchanan, C., Whitney, J., Fisher, J., Hatcher, H.C., D'Agostino Jr, R., Kock, N.D., Ajayan, P.M., Carroll, D.L., Akman, S., Torti, F.M., Torti, S.V.: Long-term survival following a single treatment of kidney tumors with multiwalled carbon nanotubes and near-infrared radiation. Proc. Natl. Acad. Sci. U S A **106**, 12897–12902 (2009). doi:10.1073/pnas.0905195106
10. Burke, A.R., Singh, R.N., Carroll, D.L., Owen, J.D., Kock, N.D., D'Agostino Jr, R., Torti, F.M., Torti, S.V.: Determinants of the thrombogenic potential of multiwalled carbon nanotubes. Biomaterials **32**, 5970–5978 (2011). doi:10.1016/j.biomaterials.2011.04.059
11. Cai, D., Mataraza, J.M., Qin, Z.H., Huang, Z., Huang, J., Chiles, T.C., Carnahan, D., Kempa, K., Ren, Z.: Highly efficient molecular delivery into mammalian cells using carbon nanotube spearing. Nat. Methods **2**, 449–454 (2005). doi:10.1038/nmeth761
12. Chakravarty, P., Marches, R., Zimmerman, N.S., Swafford, A.D., Bajaj, P., Musselman, I.H., Pantano, P., Draper, R.K., Vitetta, E.S.: Thermal ablation of tumor cells with antibody-functionalized single-walled carbon nanotubes. Proc. Natl. Acad. Sci. U S A **105**, 8697–8702 (2008). doi:10.1073/pnas.0803557105

13. Cherukuri, P., Gannon, C.J., Leeuw, T.K., Schmidt, H.K., Smalley, R.E., Curley, S.A., Weisman, R.B.: Mammalian pharmacokinetics of carbon nanotubes using intrinsic near-infrared fluorescence. Proc. Natl. Acad. Sci. U S A **103**, 18882–18886 (2006). doi:10.1073/pnas.0609265103

14. Cheung, W., Pontoriero, F., Taratula, O., Chen, A.M., He, H.: DNA and carbon nanotubes as medicine. Adv. Drug Deliv. Rev. **62**, 633–649 (2010). doi:10.1016/j.addr.2010.03.007

15. De la Zerda, A., Zavaleta, C., Keren, S., Vaithilingam, S., Bodapati, S., Liu, Z., Levi, J., Smith, B.R., Ma, T.J., Oralkan, O., Cheng, Z., Chen, X., Dai, H., Khuri-Yakub, B.T., Gambhir, S.S.: Carbon nanotubes as photoacoustic molecular imaging agents in living mice. Nat. Nanotechnol. **3**, 557–562 (2008). doi:10.1038/nnano.2008.231

16. Dhar, S., Liu, Z., Thomale, J., Dai, H., Lippard, S.J.: Targeted single-wall carbon nanotube-mediated Pt(IV) prodrug delivery using folate as a homing device. J. Am. Chem. Soc. **130**, 11467–11476 (2008). doi:10.1021/ja803036e

17. Di Crescenzo, A., Velluto, D., Hubbell, J.A., Fontana, A.: Biocompatible dispersions of carbon nanotubes: a potential tool for intracellular transport of anticancer drugs. Nanoscale **3**, 925–928 (2011). doi:10.1039/c0nr00444h

18. Dumortier, H., Lacotte, S., Pastorin, G., Marega, R., Wu, W., Bonifazi, D., Briand, J.P., Prato, M., Muller, S., Bianco, A.: Functionalized carbon nanotubes are non-cytotoxic and preserve the functionality of primary immune cells. Nano. Lett. **6**, 1522–1528 (2006). doi:10.1021/nl061160x

19. Firme III, C.P., Bandaru, P.R.: Toxicity issues in the application of carbon nanotubes to biological systems. Nanomedicine **6**, 245–256 (2010). doi:10.1016/j.nano.2009.07.003

20. Foldvari, M., Bagonluri, M.: Carbon nanotubes as functional excipients for nanomedicines: II drug delivery and biocompatibility issues. Nanomedicine **4**, 183–200 (2008). doi:10.1016/j.nano.2008.04.003

21. Georgakilas, V., Kordatos, K., Prato, M., Guldi, D.M., Holzinger, M., Hirsch, A.: Organic functionalization of carbon nanotubes. J. Am. Chem. Soc. **124**, 760–761 (2002). doi:10.1021/ja016954m

22. Hampel, S., Kunze, D., Haase, D., Kramer, K., Rauschenbach, M., Ritschel, M., Leonhardt, A., Thomas, J., Oswald, S., Hoffmann, V., Buchner, B.: Carbon nanotubes filled with a chemotherapeutic agent: a nanocarrier mediates inhibition of tumor cell growth. Nanomedicine (Lond) **3**, 175–182 (2008). doi:10.2217/17435889.3.2.175

23. Heister, E., Lamprecht, C., Neves, V., Tilmaciu, C., Datas, L., Flahaut, E., Soula, B., Hinterdorfer, P., Coley, H.M., Silva, S.R., McFadden, J.: Higher dispersion efficacy of functionalized carbon nanotubes in chemical and biological environments. ACS Nano **4**, 2615–2626 (2010). doi:10.1021/nn100069k

24. Hilder, T.A., Hill, J.M.: Modeling the loading and unloading of drugs into nanotubes. Small **5**, 300–308 (2009). doi:10.1002/smll.200800321

25. Huang, H., Yuan, Q., Shah, J.S., Misra, R.D.: A new family of folate-decorated and carbon nanotube-mediated drug delivery system: synthesis and drug delivery response. Adv. Drug Deliv. Rev. (2011). doi:10.1016/j.addr.2011.04.001

26. Huang, H.C., Barua, S., Sharma, G., Dey, S.K., Rege, K.: Inorganic nanoparticles for cancer imaging and therapy. J Control Release (2011). doi:10.1016/j.jconrel.2011.07.005

27. Iancu, C., Mocan, L., Bele, C., Orza, A.I., Tabaran, F.A., Catoi, C., Stiufiuc, R., Stir, A., Matea, C., Iancu, D., Agoston-Coldea, L., Zaharie, F., Mocan, T.: Enhanced laser thermal ablation for the in vitro treatment of liver cancer by specific delivery of multiwalled carbon nanotubes functionalized with human serum albumin. Int. J. Nanomedicine **6**, 129–141 (2011). doi:10.2147/IJN.S15841

28. Jain, K.K.: Advances in the field of nanooncology. BMC Med. **8**, 83 (2010). doi:10.1186/1741-7015-8-83

29. Ji, S.R., Liu, C., Zhang, B., Yang, F., Xu, J., Long, J., Jin, C., Fu, D.L., Ni, Q.X., Yu, X.J.: Carbon nanotubes in cancer diagnosis and therapy. Biochim. Biophys. Acta **1806**, 29–35 (2010). doi:10.1016/j.bbcan.2010.02.004

30. Kam, N.W., Liu, Z., Dai, H.: Carbon nanotubes as intracellular transporters for proteins and DNA: an investigation of the uptake mechanism and pathway. Angew. Chem. Int. Ed. Engl. **45**, 577–581 (2006). doi:10.1002/anie.200503389

31. Kam, N.W., O'Connell, M., Wisdom, J.A., Dai, H.: Carbon nanotubes as multifunctional biological transporters and near-infrared agents for selective cancer cell destruction. Proc. Natl. Acad. Sci. U S A **102**, 11600–11605 (2005). doi:10.1073/pnas.0502680102

32. Karmakar, A., Bratton, S.M., Dervishi, E., Ghosh, A., Mahmood, M., Xu, Y., Saeed, L.M., Mustafa, T., Casciano, D., Radominska-Pandya, A., Biris, A.S.: Ethylenediamine functionalized-single-walled nanotube (f-SWNT)-assisted in vitro delivery of the oncogene suppressor p53 gene to breast cancer MCF-7 cells. Int J Nanomedicine **6**, 1045–1055 (2011). doi:10.2147/IJN.S17684

33. Kostarelos, K., Bianco, A., Prato, M.: Promises, facts and challenges for carbon nanotubes in imaging and therapeutics. Nat. Nanotechnol. **4**, 627–633 (2009). doi:10.1038/nnano.2009.241

34. Kostarelos, K., Lacerda, L., Pastorin, G., Wu, W., Wieckowski, S., Luangsivilay, J., Godefroy, S., Pantarotto, D., Briand, J.P., Muller, S., Prato, M., Bianco, A.: Cellular uptake of functionalized carbon nanotubes is independent of functional group and cell type. Nat. Nanotechnol. **2**, 108–113 (2007). doi:10.1038/nnano.2006.209

35. Lay, C.L., Liu, H.Q., Tan, H.R., Liu, Y.: Delivery of paclitaxel by physically loading onto poly(ethylene glycol) (PEG)-graft-carbon nanotubes for potent cancer therapeutics. Nanotechnology **21**, 065101 (2010). doi:10.1088/0957-4484/21/6/065101

36. Li, R., Wu, R., Zhao, L., Wu, M., Yang, L., Zou, H.: P-glycoprotein antibody functionalized carbon nanotube overcomes the multidrug resistance of human leukemia cells. ACS Nano **4**, 1399–1408 (2010). doi:10.1021/nn9011225

37. Liang, F., Chen, B.: A review on biomedical applications of single-walled carbon nanotubes. Curr. Med. Chem. **17**, 10–24 (2010)

38. Liu, X., Tao, H., Yang, K., Zhang, S., Lee, S.T., Liu, Z.: Optimization of surface chemistry on single-walled carbon nanotubes for in vivo photothermal ablation of tumors. Biomaterials **32**, 144–151 (2011). doi:10.1016/j.biomaterials.2010.08.096

39. Liu, Z., Cai, W., He, L., Nakayama, N., Chen, K., Sun, X., Chen, X., Dai, H.: In vivo biodistribution and highly efficient tumour targeting of carbon nanotubes in mice. Nat. Nanotechnol. **2**, 47–52 (2007). doi:10.1038/nnano.2006.170

40. Liu, Z., Chen, K., Davis, C., Sherlock, S., Cao, Q., Chen, X., Dai, H.: Drug delivery with carbon nanotubes for in vivo cancer treatment. Cancer Res. **68**, 6652–6660 (2008). doi:10.1158/0008-5472.CAN-08-1468

41. Liu, Z., Davis, C., Cai, W., He, L., Chen, X., Dai, H.: Circulation and long-term fate of functionalized, biocompatible single-walled carbon nanotubes in mice probed by Raman spectroscopy. Proc. Natl. Acad. Sci. U S A **105**, 1410–1415 (2008). doi:10.1073/pnas.0707654105

42. Liu, Z., Li, X., Tabakman, S.M., Jiang, K., Fan, S., Dai, H.: Multiplexed multicolor Raman imaging of live cells with isotopically modified single walled carbon nanotubes. J. Am. Chem. Soc. **130**, 13540–13541 (2008). doi:10.1021/ja806242t

43. Liu, Z., Sun, X., Nakayama-Ratchford, N., Dai, H.: Supramolecular chemistry on water-soluble carbon nanotubes for drug loading and delivery. ACS Nano **1**, 50–56 (2007). doi:10.1021/nn700040t

44. Liu, Z., Tabakman, S., Welsher, K., Dai, H.: Carbon nanotubes in biology and medicine: in vitro and in vivo detection, imaging and drug delivery. Nano Res **2**, 85–120 (2009). doi:10.1007/s12274-009-9009-8

45. Luo, J., Solimini, N.L., Elledge, S.J.: Principles of cancer therapy: oncogene and non-oncogene addiction. Cell **136**, 823–837 (2009). doi:10.1016/j.cell.2009.02.024

46. Maeda, H., Bharate, G.Y., Daruwalla, J.: Polymeric drugs for efficient tumor-targeted drug delivery based on EPR-effect. Eur. J. Pharm. Biopharm. **71**, 409–419 (2009). doi:10.1016/j.ejpb.2008.11.010

47. Maeda, H., Wu, J., Sawa, T., Matsumura, Y., Hori, K.: Tumor vascular permeability and the EPR effect in macromolecular therapeutics: a review. J Control Release **65**, 271–284 (2000). doi:S0168-3659(99)00248-5

48. Marches, R., Mikoryak, C., Wang, R.H., Pantano, P., Draper, R.K., Vitetta, E.S.: The importance of cellular internalization of antibody-targeted carbon nanotubes in the photothermal ablation of breast cancer cells. Nanotechnology **22**, 095101 (2011). doi:10.1088/0957-4484/22/9/095101

49. Matsumura, S., Ajima, K., Yudasaka, M., Iijima, S., Shiba, K.: Dispersion of cisplatin-loaded carbon nanohorns with a conjugate comprised of an artificial peptide aptamer and polyethylene glycol. Mol. Pharm. **4**, 723–729 (2007). doi:10.1021/mp070022t

50. McDevitt, M.R., Chattopadhyay, D., Kappel, B.J., Jaggi, J.S., Schiffman, S.R., Antczak, C., Njardarson, J.T., Brentjens, R., Scheinberg, D.A.: Tumor targeting with antibody-functionalized, radiolabeled carbon nanotubes. J. Nucl. Med. **48**, 1180–1189 (2007). doi:10.2967/jnumed.106.039131

51. Misra, R., Acharya, S., Sahoo, S.K.: Cancer nanotechnology: application of nanotechnology in cancer therapy. Drug Discov Today **15**, 842–850 (2010). doi:10.1016/j.drudis.2010.08.006

52. Miyawaki, J., Yudasaka, M., Azami, T., Kubo, Y., Iijima, S.: Toxicity of single-walled carbon nanohorns. ACS Nano **2**, 213–226 (2008). doi:10.1021/nn700185t

53. Mocan, L., Tabaran, F.A., Mocan, T., Bele, C., Orza, A.I., Lucan, C., Stiufiuc, R., Manaila, I., Iulia, F., Dana, I., Zaharie, F., Osian, G., Vlad, L., Iancu, C.: Selective ex vivo photothermal ablation of human pancreatic cancer with albumin functionalized multiwalled carbon nanotubes. Int J Nanomedicine **6**, 915–928 (2011). doi:10.2147/IJN.S19013

54. Murakami, T., Ajima, K., Miyawaki, J., Yudasaka, M., Iijima, S., Shiba, K.: Drug-loaded carbon nanohorns: adsorption and release of dexamethasone in vitro. Mol. Pharm. **1**, 399–405 (2004). doi:10.1021/mp049928e

55. Murugesan, S., Park, T.J., Yang, H., Mousa, S., Linhardt, R.J.: Blood compatible carbon nanotubes–nano-based neoproteoglycans. Langmuir **22**, 3461–3463 (2006). doi:10.1021/la0534468

56. Nimmagadda, A., Thurston, K., Nollert, M.U., McFetridge, P.S.: Chemical modification of SWNT alters in vitro cell-SWNT interactions. J. Biomed. Mater. Res. A **76**, 614–625 (2006). doi:10.1002/jbm.a.30577

57. Pacurari, M., Qian, Y., Porter, D.W., Wolfarth, M., Wan, Y., Luo, D., Ding, M., Castranova, V., Guo, N.L.: Multi-walled carbon nanotube-induced gene expression in the mouse lung: association with lung pathology. Toxicol. Appl. Pharmacol. **255**, 18–31 (2011). doi:10.1016/j.taap.2011.05.012

58. Pantarotto, D., Partidos, C.D., Hoebeke, J., Brown, F., Kramer, E., Briand, J.P., Muller, S., Prato, M., Bianco, A.: Immunization with peptide-functionalized carbon nanotubes enhances virus-specific neutralizing antibody responses. Chem. Biol. **10**, 961–966 (2003). doi:S107455210300214X

59. Park, Y.K., Bold, B., Lee, W.K., Jeon, M.H., An, K.H., Jeong, S.Y., Shim, Y.K.: d-(+)-Galactose-conjugated single-walled carbon nanotubes as new chemical probes for electrochemical biosensors for the cancer marker galectin-3. Int. J. Mol. Sci. **12**, 2946–2957 (2011). doi:10.3390/ijms12052946

60. Pastorin, G., Wu, W., Wieckowski, S., Briand, J.P., Kostarelos, K., Prato, M., Bianco, A.: Double functionalization of carbon nanotubes for multimodal drug delivery. Chem. Commun. (Camb) 1182–1184 (2006). doi: 10.1039/b516309a

61. Prakash, S., Malhotra, M., Shao, W., Tomaro-Duchesneau, C., Abbasi, S.: Polymeric nanohybrids and functionalized carbon nanotubes as drug delivery carriers for cancer therapy. Adv. Drug Deliv. Rev. (2011). doi:10.1016/j.addr.2011.06.013

62. Pramanik, M., Swierczewska, M., Green, D., Sitharaman, B., Wang, L.V.: Single-walled carbon nanotubes as a multimodal-thermoacoustic and photoacoustic-contrast agent. J. Biomed. Opt. **14**, 034018 (2009). doi:10.1117/1.3147407

63. Prato, M., Kostarelos, K., Bianco, A.: Functionalized carbon nanotubes in drug design and discovery. Acc. Chem. Res. **41**, 60–68 (2008). doi:10.1021/ar700089b

64. Pulskamp, K., Diabate, S., Krug, H.F.: Carbon nanotubes show no sign of acute toxicity but induce intracellular reactive oxygen species in dependence on contaminants. Toxicol. Lett. **168**, 58–74 (2007). doi:10.1016/j.toxlet.2006.11.001

65. Raffa, V., Ciofani, G., Vittorio, O., Riggio, C., Cuschieri, A.: Physicochemical properties affecting cellular uptake of carbon nanotubes. Nanomedicine (Lond.) **5**, 89–97 (2010). doi:10.2217/nnm.09.95

66. Ruggiero, A., Villa, C.H., Holland, J.P., Sprinkle, S.R., May, C., Lewis, J.S., Scheinberg, D.A., McDevitt, M.R.: Imaging and treating tumor vasculature with targeted radiolabeled carbon nanotubes. Int. J. Nanomedicine **5**, 783–802 (2010). doi:10.2147/IJN.S13300

67. Sahoo, N.G., Bao, H., Pan, Y., Pal, M., Kakran, M., Cheng, H.K., Li, L., Tan, L.P.: Functionalized carbon nanomaterials as nanocarriers for loading and delivery of a poorly water-soluble anticancer drug: a comparative study. Chem. Commun. (Camb.) **47**, 5235–5237 (2011). doi:10.1039/c1cc00075f

68. Shvedova, A.A., Kisin, E.R., Porter, D., Schulte, P., Kagan, V.E., Fadeel, B., Castranova, V.: Mechanisms of pulmonary toxicity and medical applications of carbon nanotubes: two faces of Janus? Pharmacol. Ther. **121**, 192–204 (2009). doi:10.1016/j.pharmthera.2008.10.009

69. Singh, R., Pantarotto, D., Lacerda, L., Pastorin, G., Klumpp, C., Prato, M., Bianco, A., Kostarelos, K.: Tissue biodistribution and blood clearance rates of intravenously administered carbon nanotube radiotracers. Proc. Natl. Acad. Sci. U S A **103**, 3357–3362 (2006). doi:10.1073/pnas.0509009103

70. Singh, R., Pantarotto, D., McCarthy, D., Chaloin, O., Hoebeke, J., Partidos, C.D., Briand, J.P., Prato, M., Bianco, A., Kostarelos, K.: Binding and condensation of plasmid DNA onto functionalized carbon nanotubes: toward the construction of nanotube-based gene delivery vectors. J. Am. Chem. Soc. **127**, 4388–4396 (2005). doi:10.1021/ja0441561

71. Sobhani, Z., Dinarvand, R., Atyabi, F., Ghahremani, M., Adeli, M.: Increased paclitaxel cytotoxicity against cancer cell lines using a novel functionalized carbon nanotube. Int. J. Nanomedicine **6**, 705–719 (2011). doi:10.2147/IJN.S17336

72. Tasis, D., Tagmatarchis, N., Bianco, A., Prato, M.: Chemistry of carbon nanotubes. Chem. Rev. **106**, 1105–1136 (2006). doi:10.1021/cr050569o

73. Tasis, D., Tagmatarchis, N., Georgakilas, V., Prato, M.: Soluble carbon nanotubes. Chemistry **9**, 4000–4008 (2003). doi:10.1002/chem.200304800

74. Thomas, L.V., Arun, U., Remya, S., Nair, P.D.: A biodegradable and biocompatible PVA-citric acid polyester with potential applications as matrix for vascular tissue engineering. J. Mater. Sci. Mater. Med. **20**(Suppl 1), S259–S269 (2009). doi:10.1007/s10856-008-3599-7

75. Valcarcel, M., Cardenas, S., Simonet, B.M.: Role of carbon nanotubes in analytical science. Anal. Chem. **79**, 4788–4797 (2007). doi:10.1021/ac070196m

76. Venturelli, E., Fabbro, C., Chaloin, O., Menard-Moyon, C., Smulski, C.R., Da, R.T., Kostarelos, K., Prato, M., Bianco, A.: Antibody covalent immobilization on carbon nanotubes and assessment of antigen binding. Small **7**, 2179–2187 (2011). doi:10.1002/smll.201100137

77. Villa, C.H., Dao, T., Ahearn, I., Fehrenbacher, N., Casey, E., Rey, D.A., Korontsvit, T., Zakhaleva, V., Batt, C.A., Philips, M.R., Scheinberg, D.A.: Single-walled carbon nanotubes deliver peptide antigen into dendritic cells and enhance IgG responses to tumor-associated antigens. ACS Nano **5**, 5300–5311 (2011). doi:10.1021/nn200182x

78. Vittorio, O., Raffa, V., Cuschieri, A.: Influence of purity and surface oxidation on cytotoxicity of multiwalled carbon nanotubes with human neuroblastoma cells. Nanomedicine **5**, 424–431 (2009). doi:10.1016/j.nano.2009.02.006

79. Wang, J., Liu, G., Jan, M.R.: Ultrasensitive electrical biosensing of proteins and DNA: carbon-nanotube derived amplification of the recognition and transduction events. J. Am. Chem. Soc. **126**, 3010–3011 (2004). doi:10.1021/ja031723w

80. Wang, L., Luanpitpong, S., Castranova, V., Tse, W., Lu, Y., Pongrakhananon, V., Rojanasakul, Y.: Carbon nanotubes induce malignant transformation and tumorigenesis of human lung epithelial cells. Nano Lett. **11**, 2796–2803 (2011). doi:10.1021/nl2011214

81. Wang, X., Jia, G., Wang, H., Nie, H., Yan, L., Deng, X.Y., Wang, S.: Diameter effects on cytotoxicity of multi-walled carbon nanotubes. J. Nanosci. Nanotechnol. **9**, 3025–3033 (2009)
82. Warheit, D.B., Laurence, B.R., Reed, K.L., Roach, D.H., Reynolds, G.A., Webb, T.R.: Comparative pulmonary toxicity assessment of single-wall carbon nanotubes in rats. Toxicol. Sci. **77**, 117–125 (2004). doi:10.1093/toxsci/kfg228
83. Welsher, K., Liu, Z., Daranciang, D., Dai, H.: Selective probing and imaging of cells with single walled carbon nanotubes as near-infrared fluorescent molecules. Nano Lett. **8**, 586–590 (2008). doi:10.1021/nl072949q
84. Whitney, J.R., Sarkar, S., Zhang, J., Do, T., Young, T., Manson, M.K., Campbell, T.A., Puretzky, A.A., Rouleau, C.M., More, K.L., Geohegan, D.B., Rylander, C.G., Dorn, H.C., Rylander, M.N.: Single walled carbon nanohorns as photothermal cancer agents. Lasers Surg. Med. **43**, 43–51 (2011). doi:10.1002/lsm.21025
85. Worle-Knirsch, J.M., Pulskamp, K., Krug, H.F.: Oops they did it again! Carbon nanotubes hoax scientists in viability assays. Nano Lett. **6**, 1261–1268 (2006). doi:10.1021/nl060177c
86. Wu, H., Liu, G., Wang, X., Zhang, J., Chen, Y., Shi, J., Yang, H., Hu, H., Yang, S.: Solvothermal synthesis of cobalt ferrite nanoparticles loaded on multiwalled carbon nanotubes for magnetic resonance imaging and drug delivery. Acta Biomater. **7**, 3496–3504 (2011). doi:10.1016/j.actbio.2011.05.031
87. Wu, W., Li, R., Bian, X., Zhu, Z., Ding, D., Li, X., Jia, Z., Jiang, X., Hu, Y.: Covalently combining carbon nanotubes with anticancer agent: preparation and antitumor activity. ACS Nano **3**, 2740–2750 (2009). doi:10.1021/nn9005686
88. Wu, Y., Phillips, J.A., Liu, H., Yang, R., Tan, W.: Carbon nanotubes protect DNA strands during cellular delivery. ACS Nano **2**, 2023–2028 (2008). doi:10.1021/nn800325a
89. Yang, F., Hu, J., Yang, D., Long, J., Luo, G., Jin, C., Yu, X., Xu, J., Wang, C., Ni, Q., Fu, D.: Pilot study of targeting magnetic carbon nanotubes to lymph nodes. Nanomedicine (Lond.) **4**, 317–330 (2009). doi:10.2217/nnm.09.5
90. Yang, F., Jin, C., Yang, D., Jiang, Y., Li, J., Di, Y., Hu, J., Wang, C., Ni, Q., Fu, D.: Magnetic functionalised carbon nanotubes as drug vehicles for cancer lymph node metastasis treatment. Eur. J. Cancer **47**, 1873–1882 (2011). doi:10.1016/j.ejca.2011.03.018
91. Zhang, H., Jiang, H., Sun, F., Wang, H., Zhao, J., Chen, B., Wangb, X.: Rapid diagnosis of multidrug resistance in cancer by electrochemical sensor based on carbon nanotubes–drug supramolecular nanocomposites. Biosens. Bioelectron. **26**, 3361–3366 (2011). doi:10.1016/j.bios.2011.01.020
92. Zhang, X., Meng, L., Lu, Q., Fei, Z., Dyson, P.J.: Targeted delivery and controlled release of doxorubicin to cancer cells using modified single wall carbon nanotubes. Biomaterials **30**, 6041–6047 (2009). doi:10.1016/j.biomaterials.2009.07.025
93. Zhang, Y., Bai, Y., Yan, B.: Functionalized carbon nanotubes for potential medicinal applications. Drug Discov. Today **15**, 428–435 (2010). doi:10.1016/j.drudis.2010.04.005
94. Zhou, F., Xing, D., Ou, Z., Wu, B., Resasco, D.E., Chen, W.R.: Cancer photothermal therapy in the near-infrared region by using single-walled carbon nanotubes. J. Biomed. Opt. **14**, 021009 (2009). doi:10.1117/1.3078803
95. Zhou, F., Xing, D., Wu, B., Wu, S., Ou, Z., Chen, W.R.: New insights of transmembranal mechanism and subcellular localization of noncovalently modified single-walled carbon nanotubes. Nano Lett. **10**, 1677–1681 (2010). doi:10.1021/nl100004m

CNTs/TiO$_2$ Composites

Silvana Da Dalt, Annelise Kopp Alves and C. P. Bergmann

Abstract Titanium dioxide (TiO$_2$) is a semiconductor material that is widely used in many different areas, such as gas sensors, air purification, catalysis, solar to electric energy conversion, photoelectrochemical systems and photocatalyst for degrading a wide range of organic pollutants because of its nontoxicity, photochemical stability, and low cost. There are reports that show that the heterojunction of TiO$_2$ and carbon nanotubes (CNTs) improves the efficiency of the photocatalytic activity, mainly because the recombination of the photogenerated electron–hole pairs becomes more difficult in the presence of nanotubes. Multiwall carbon nanotubes/TiO$_2$ (MWCNT/TiO$_2$) composite materials have been attracting attention in relation to their use in the treatment of contaminated water and air by heterogeneous photocatalysis, hydrogen evolution, CO$_2$ photo-reduction, and dye sensitized solar cells. Nevertheless, functionalization routes to aggregate these materials and characterization methods need to be studied; since they have direct influence on properties and potential applications.

Abbreviations

AFM	Atomic Force Microscopy
BET	Brunauer–Emmett–Teller
CNTs	Carbon Nanotubes
EDX	Energy Dispersive X Ray
FESEM	Field Emission Scanning Electron Microscopy
FT-IR	Fourier Transformer Infrared

S. Da Dalt (✉)
Av. Osvaldo, Aranha 99 - 705C, Porto Alegre, 90035-190, Brazil
e-mail: silvana.da.dalt@ufrgs.br

S. Da Dalt · A. K. Alves · C. P. Bergmann
Escola de Engenharia, Universidade Federal do Rio Grande do Sul, Porto Alegre, 90035-190, Brazil

C. Avellaneda (ed.), *NanoCarbon 2011*, Carbon Nanostructures, DOI: 10.1007/978-3-642-31960-0_6, © Springer-Verlag Berlin Heidelberg 2013

HPO   Heterogeneous Photocatalytic Oxidation
MB   Methylene Blue
MO   Methyl Orange
MWCNT   Multi-Walled Carbon Nanotubes
SWCNT   Single-Walled Carbon Nanotubes
SEM   Scanning Electron Microscopy
TEM   Transmission Electron Microscopy
TGA   Thermo Gravimetric Analysis
UV   Ultraviolet
XRD   X Ray Diffraction

1 Introduction

During the past 10 years, since the first study of CNTs coatings on oxide layers by Seeger et al. [1], composites containing CNTs are believed to have many different applications and exhibit combination effects between carbon phases and metal oxides. The progress in research has allowed CNTs to be used as catalyst supports as an alternative to activated carbon due to the combination of their electronic, adsorption, mechanical and thermal properties. Because CNTs can conduct electrons, have high surface areas and high adsorption capacities, they are being employed as supports for TiO_2 to be used as photocatalysts [2]. In this review, we analyze investigations on the functionalization of CNTs with TiO_2. Our studies were based on researches focused on the technological applications, fundamentals and properties of CNTs and TiO_2, as well as by the discussions on synthesis and characterization of these composites.

1.1 Carbon Nanotubes

Since the discovery of fullerenes [3] carbon nanostructures have been the focus of attention. The sp^2 geometrical based carbon structure can result in new symmetries and structures, and consequently extraordinary new properties. CNTs represent the most striking example of this advent. The results show basically that CNTs behave like rolled-up cylinders of graphene sheets of sp^2 bonded carbon atoms, except that the tubule diameters are small enough to exhibit the effects of one-dimensional (1D) periodicity [4]. Two decades after their discovery [5], studies and outcomes of these nanostructures indicate that each year a growing number of practical applications are investigated. There have been great improvements in synthesis techniques, which can produce pure nanotubes in considerable quantities. Studies of structure and property relations in nanotubes have been strongly supported by

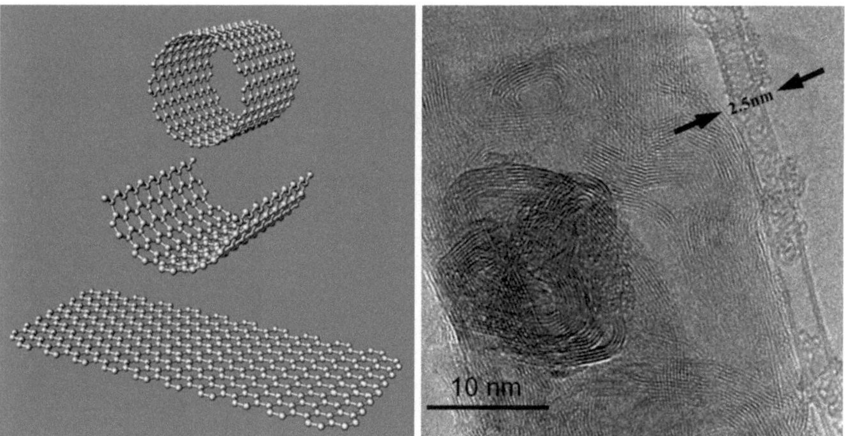

Fig. 1 The structure of single-walled (SWCNT) formed by wrapping a single sheet of graphene (multi-walled (MWCNT) is obtained from two or more sheets of graphene) (**a**). TEM image of MWCNTs with diameter of approximately 2.5 nm (**b**)

theoretical modeling that has provided understanding for experiments and thus, the rapid expansion of this field [6].

CNTs may consist of one graphene sheet (single-walled CNTs, SWCNTs) or several graphene sheets (multi-walled CNTs, MWCNTs) creating cylinders concentrically stacked with an adjacent layer spacing of ~ 0.35 nm (Fig. 1). Because of the symmetry and unique electronic structure of graphene, the structure of a SWCNT determines its electrical properties. The electronic properties of CNTs determine if they are metallic or semiconducting, depending on their geometry [7].

The first evidence of carbon nanotubes may have been reported by Hillert and Lange [8] who found carbon filaments with inner cavities, however this scientific production was entitled based on the study of graphite nanofibers. The interest in fibrous carbon has since then been recurrent in research of the carbon nanostructure field, and coincides with the discovery of MWCNT by Iijima [5] and is followed by the production of single-wall carbon nanotubes [9]. Since then, CNTs have become one of the most active fields of Nanoscience and Nanotechnology [10], mainly due to their potential applications as a support for dispersion of materials in order to give them additional functional properties, such as structure, surface area, activity, and conductivity. Inorganic species are involved in these applications, such as inorganic oxide coatings, especially semiconductors and insulators, although the still uncontrollable nature of the surface reactions of CNTs with inorganic oxides is in some cases a challenge [11].

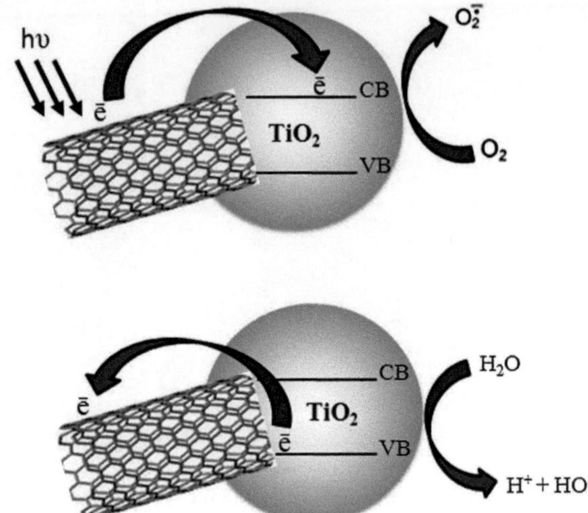

Fig. 2 Processes in semiconductor photocatalysis. Photo-absorption and electron–hole pair generation from change of separation and migration of the electron to the surface reaction

1.2 Titanium Dioxide and Photocatalysis

Titanium dioxide occurs naturally on the Earth's surface as three mineral compounds known as anatase, brookite, and rutile. It is usually extracted from titanium tetrachloride by carbon reduction and re-oxidization, but it may also be processed from ilmenite by reduction with sulfuric acid. One of the most common commercial nanoscale TiO_2 powders is made by Evonik via high-temperature flame hydrolysis and known as AEROXIDE® TiO_2 P25 [12].

One of the most popular uses of TiO_2 is as photocatalyst. A photocatalyst decreases the activation energy of photoinduced processes. In summary, photocatalysis is a process where a chemical reaction is accelerated in presence of a catalyst that is activated in the presence of light. Utilization of the photocatalytic effect is already seen in applications such as self-cleaning tiles, windows and textiles, anti-fogging car mirrors, and anti-microbial coatings [13–16].

A photocatalytic system consists of a semiconductor in contact with a liquid or a gaseous medium and an illumination method. The band gap value of anatase is 3.2 eV. Light of a wavelength lower than 385 nm can therefore excite electrons from the valence band to the conduction band, producing an electron–hole pair. If the system contains water and dissolved oxygen, water gets oxidized by the positive holes and it splits into $\cdot$OH and H^+ and, since oxygen is an easily reducible substance, different types of reactive oxygen species such as H_2O_2 and O^{-2} are generated (Fig. 2).

These oxidizing agents are able to decompose organic compounds into CO_2 and H_2O, which is useful in air, effluent and water purification. In fact, due to its ability to produce free radicals when it is illuminated with the proper wavelength, TiO_2 has applications in architecture, the automotive, textile and glass industries, and also in environmental protection [17–23].

Another interesting application of TiO$_2$ that uses this free radical formation is microbiological disinfection. In a photosterilization process the free radicals attack the cells of microorganisms, so that the growth of bacteria, viruses, algae, yeast, mold, and other microorganisms on surfaces or in liquids is effectively inhibited [24–28].

A major challenge in photocatalysis research is to improve the activity of the catalysts either by making it work faster or by making it active in visible light (sun light).

One way to enhance the activity of the photocatalyst and make it work faster is to use nanoparticles which have a high surface area and, possibly, to improve size quantization effects. The former characteristic exposes more area to UV illumination, making the reaction occur faster because more free radicals can be formed. And the latter process produces an increase in the absorption coefficient at specific wavelengths [29].

Another way to enhance the activity of the photocatalyst is to shift the optical response of the TiO$_2$ from the UV to the visible spectral range. Since UV light accounts for less than 10 % of the sun's energy compared to visible light (45 %), any shift in its optical response will have a positive effect on the photocatalytic efficiency of the material in natural conditions. This optical shift can usually be achieved by doping TiO$_2$ with other elements. Several studies have doped TiO$_2$ with different elements, such as Cs, Ag, Fe, Co [30], N-SiO$_2$ [31], fluorine-doped TiO$_2$ [32], N-doped, F-doped, and N–F-co-doped TiO$_2$ powders [33], among many others. Nowadays, researchers have focused their work to promote faster photocatalytic reactions in UV and also in visible light using pure carbon nanotubes (CNT) or using doped titanium oxide with carbon nanotubes.

1.3 Application of CNT/TiO$_2$ Composites

TiO$_2$ has been studied for a long time for the oxidation or reduction of inorganic and organic species in air and water. In air treatment, for example, it is possible to apply TiO$_2$ to eliminate smells, thus deodorizing the ambient. The elimination of toxic chemicals from wastewater is currently one of the most crucial subjects in pollution control. The large amount of dyes used in the dyeing stage of textile manufacturing processes represents an increasing environmental danger due to the carcinogenic nature of these dyes [34]. Other applications for CNTs/TiO$_2$ are possible, as exemplified in Fig. 3. The efficiency of the photocatalytic activity is affected by the recombination of the photogenerated electron hole pairs, so a lot of effort has been put on the coupling of TiO$_2$ with organic and inorganic materials.

In the work of Luo et al. [35]. CNTs were obtained using the CVD technique. After purification with acids, washing with deionized water, milling and annealing at 1200 °C under a high vacuum atmosphere, highly defective CNTs were obtained. It is known that the presence of defects in the structure of the photocatalyst, especially oxygen vacancies, promotes photo-reactions. By creating

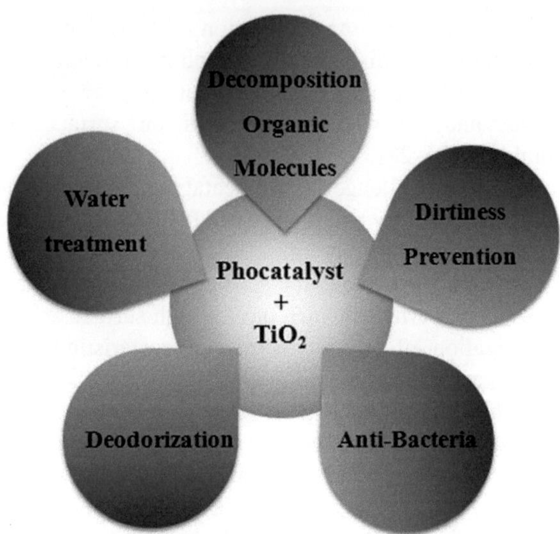

Fig. 3 Some applications of photocatalytic properties of TiO_2

defective CNTs, Luo et al. [35], demonstrated the photocatalytic efficiency of this material in pure form under visible light for the photodegradation of H_2O_2.

Gao et al. [36] obtained a hybrid catalyst composed of CNT/TiO_2 prepared via the sol–gel method, using tetra-*n*-butyl titanate as precursor and 10 % wt of carbon nanotubes. After sintering between 350–500 °C, the anatase phase was detected using the XRD technique. The photocatalysis test was carried out using visible light and a solution of acridine dye and the composites were previously stirred overnight in the dark. The most photoactive composite was the sample sintered at 300 °C. Even though the formation of the anatase phase was incomplete at this temperature, the AFM and TGA results show that this sample had a net-like morphology with well distributed TiO_2 and CNTs over the surface.

Hu et al. [37] have reported the preparation and the characterization of the photocatalytic properties of hybrid nanofibers/mats of anatase and multi-walled carbon nanotubes (MWCNTs) using a combination of sol–gel and electrospinning techniques. They used a mass ratio of TiO_2:MWCNTs = 100:20, and for the electrospinning production of titania fibers, they used tetra-*n*-butyl titanate as precursor. For the photocatalytic characterization, methyl orange (MO) was chosen as the model pollutant and its degradation was done under UV irradiation. Before the photocatalytic tests began, the suspension containing photocatalyst nanofibrous mats and MO was kept overnight in the dark to establish adsorption–desorption equilibrium.

Wang et al. [38] prepared nanocomposites of TiO_2 nanoparticles with MWCNTs using the nanocoating-hydrothermal method. They found that the TiO_2 nanoparticles were firmly anchored on the MWCNTs through ester bonds. The photocatalytic test was done using methylene blue (MB) under visible light irradiation and the results were very interesting.

In summary, CNTs and composites of TiO$_2$ with moderate content of CNTs, exhibit excellent photocatalytic activity. The improvement of photocatalysis efficiency probably comes from two main factors: first, because of the presence of strong chemical bonding between TiO$_2$ and CNTs, there is a decrease in the agglomeration of TiO$_2$ nanoparticles and consequently an increase in the surface area and reaction sites. Second, CNTs can quickly transport charges generated during the photocatalysis process, minimizing the chance of e$^-$/h$^+$ pair recombination. It is therefore believed that the benefits of coupling CNTs and TiO$_2$ reflect potential and real applications that can be explored.

2 Synthesis Routes of CNTs/TiO$_2$ Composites

Functionalized CNTs have been studied in the last few years with great interest, due to the countless potential applications of these materials. In general, the functionalization methods could be divided into two categories: covalent and non-covalent functionalizations. Covalent functionalization is established by the formation of a covalent bond between various functional groups and the wall of the nanotube. Non-covalent functionalization is mainly based on supramolecular complexation using various adsorption forces, such as the Van der Waals force, hydrogen bonds and the electrostatic force [39].

In short, the functionalization of CNTs involves the production of defects or a change from the sp^2 to sp^3 structure, which makes the adsorption of functional groups (hydroxyl, carboxyl, ester, among others) at the outer wall of the CNT easier to occur. However, metals [40, 41], polymers [42–45], biomolecules [46, 47], organic dyes [48] and ceramic oxides [11–49] can also be adhered to these walls.

Figure 4 shows the numbers of published items for each year, using the search terms involving structures and functionalized carbon/TiO$_2$ composites. From these figures it is possible to observe that the researches involving functionalization of CNTs have intensified each year, increasing in scale, almost linearly. This could be associated with the increasing search for new properties and applications, preceded by new methods of synthesis or mixing methods. Research on structures of carbon and TiO$_2$, however, seems to have reached its peak at the end of the last decade, and this may be associated with limited applications and difficulties in obtaining CNTs.

The success of coating the carbon nanotubes surface with TiO$_2$ depends on the coating technique. The uniformity of the oxide coating varies according to the preparation method. Inorganic species are involved in these applications, especially semiconductors, but are still quite limited. This might be due to the uncontrollable nature of the surface reactions of CNTs with inorganic oxides [11].

CNTs/TiO$_2$ composites can be obtained through different methods, including mechanical mixing of TiO$_2$ and CNTs, sol–gel synthesis, electrospinning methods, chemical vapor deposition, and others. The electrospinning method is not a trivial

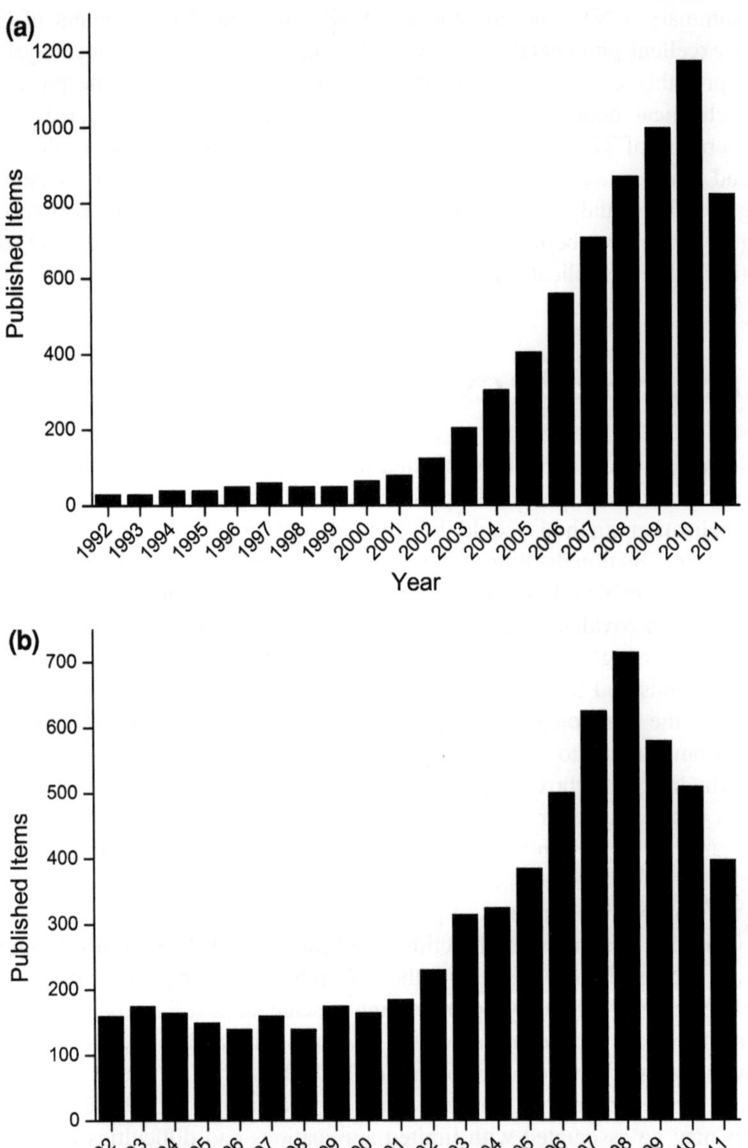

Fig. 4 The graphs show the published items in each year on functionalized carbon structures (**a**) and TiO_2 on carbon structures (**b**). The searched terms were "carbon and functionalization" (**a**) and "carbon and TiO_2" (**b**). Source: Web of Knowledge accessed on 09/05/2011

technique, requiring specialized equipment and generally it is not easy to quantify the ratio between the composite compounds. The sol–gel method is simpler, but in some cases it produces a non-uniform coating of CNTs by TiO$_2$ [50].

Jitianu et al. [51] report the TiO$_2$ coating of carbon MWCNTs by a sol–gel method using classical alkoxides like Ti(OEt)$_4$ and Ti(OPri)$_4$ and by hydrothermal hydrolysis of TiOSO$_4$ in sulfuric acid under elevated pressure. Through the sol–gel method, the MWCNTs surface was continuously coated with TiO$_2$ after thermal treatment. Although the hydrothermal method results in a good cover of the surface, the nanotubes surface is partially damaged due to the oxidizing medium of deposition. In this case, the sol–gel method is more efficient for TiO$_2$ coating on MWCNT.

The electrospinning process is capable of generating nanofibers from various polymers or inorganic/organic hybrid nanocomposites or inorganic precursors by application of an electrostatic force. CNTs/TiO$_2$ composite nanofibers are prepared by sol–gel processing followed by the electrospinning technique. The research uses organometal alkoxides along with carbon nanotubes, which can then undergo in situ stabilization by scavenging polymers, which renders the formation of continuous nanofibers [52].

CNTs/TiO$_2$ were produced as thin films on glass slides by using the so-called doctor blade method by Sampaio et al. [53]. This method can be easily employed as a fast and low-energy technique for mass production of thin films with good uniformity and reproducible properties.

MWCNTs based, metal oxide composites can also be prepared by means of hydrolysis. The successful fabrication of CNT/TiO$_2$ composites can be explained by the presence of some polar oxygenated groups such as C–O, C=O and O–C=O, which might stimulate formation of the composites, and enhance the interfacial combination of TiO$_2$ with carbon nanotubes. The fabrication of CNT/TiO$_2$ composites was carried out from CNTs dispersed into an acetone solution of Ti(OBu)$_4$ by ultrasonication. Sequentially deionized water was added to promote the hydrolysis in the mixture and, finally, CNT/TiO$_2$ composites could be obtained after heating in a vacuum oven at 150 °C for at least 8 h [54].

According to Jiang et al. [55] MWCNTs/TiO$_2$ nanocomposites were obtained using a liquid phase deposition method, that could be operated under mild reaction conditions (at 60 °C and little more than 1 atm). The anatase phase was grown directly on the MWCNTs support. A high specific surface area and uniform crystallite size was therefore achieved by this wet chemical process.

The goal of the studies listed above was, in general, to evaluate the performance in photocatalytic oxidation reactions in aqueous media. This kind of analysis is present mostly the papers, due his enlarged field the applications.

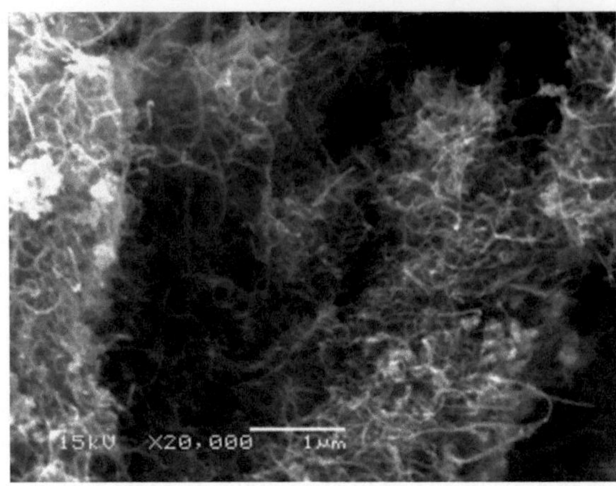

Fig. 5 SEM image of MWCNTs/TiO$_2$ composites synthesized from TiO$_2$-P25 and Baytubes® C 150 P

3 Characterization CNTs/TiO$_2$

The characterization of CNT/TiO$_2$ composites is not an easy practice, due to the physical and chemical circumstances involving such composites. The verification of the particle size of the oxide that covers the nanotube and the distance of the bond established between the CNT and TiO$_2$ are some parameters that can be analyzed by transmission electron microscopy (TEM).

The infrared spectroscopy (FT-IR) exhibits all the modifications of the CNTs structure and reveals the nature of compounds added to the CNTs [56], enabling the determination of the nature of some of the bonds between Ti and C.

The analysis of the coating physics characteristics is common in studies about functionalization of CNT with TiO$_2$, generally using scanning electron microscopy (MWCNT) (Fig. 5).

Another way to identify elements and the degree of defects in the CNT structure is through Raman spectroscopy, from which one can get important information, especially about the bond type between the nanotube and TiO$_2$. This technique allows visualization of specific and well- defined bands for both materials involved and can therefore be widely used.

Using specific surface area tests, it is possible to compare the values between TiO$_2$ and nanotubes. Usually, MWCNT/TiO$_2$ composites have a higher degradation rate than pure TiO$_2$. CNTs have relatively high surface areas, and when combined with TiO$_2$, there is usually an increase in the value of the surface area of the oxide, which leads to their higher adsorptive capacity [55] and thus, higher photoactivity.

From the thermo gravimetric analysis (TGA) it is possible to evaluate the decomposition temperatures of CNT oxide composites. It is known that the oxidation temperature of CNTs is between 600 and 650 °C [57], and in this case the TiO$_2$ is more stable at high temperatures when compared to CNTs.

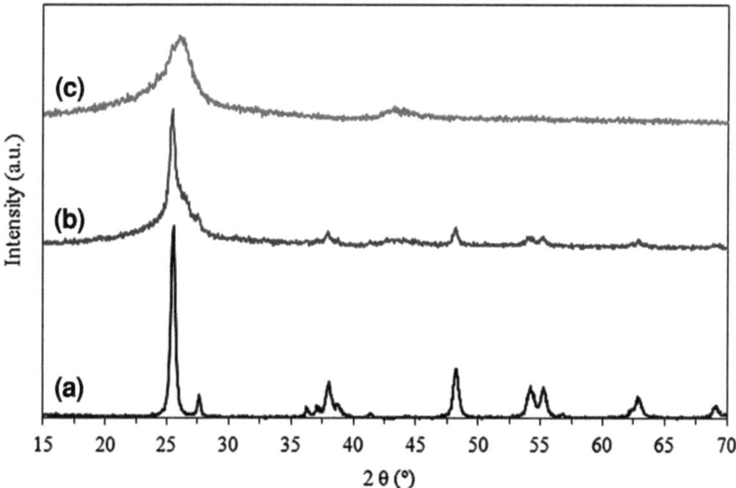

Fig. 6 X Ray Diffraction patterns of materials employed to obtain nanocarbon composites. TiO$_2$ pure (AEROXIDE®-P25) (a), MWCNTs recovered by TiO$_2$ (b) and MWCNTs pure (Baytubes® C 150 P)

The characteristic X ray diffraction (XRD) peaks of pure MWCNTs are located at $2\theta = 26°$ and $2\theta = 43.4°$, as identified in Fig. 6c. XRD peaks from the MWCNT/TiO$_2$ composites (Fig. 6b) become difficult to analyze because together with the main CNT peak at $2\theta = 26°$ [55–58] there is an interference deriving from the main peak of the anatase phase ($2\theta = 25.4°$) [59, 60]. It is observed that the peak widths of pure TiO$_2$ (Fig. 6a) widen slightly in the presence of MWCNTs. The presence of MWCNTs in the composite is responsible for the homogeneous dispersion of the TiO$_2$ matrix and, generally, this conclusion can be supported by SEM and strengthened with the analysis of broad peaks in their XRD patterns [2–61].

The influence of MWCNTs on the photocatalysis reaction is as electron acceptor because of the semiconductor nature of this material [55]. An improvement can therefore be expected in the photocatalytic performance of TiO$_2$ composites. One of the most important characterization procedures for materials used in photocatalytic applications consists in the determination of the optical absorption spectrum of a certain dye, used as a model to follow its photo-decomposition [53].

Table 1 summarizes the most common analytical techniques employed in the characterization of CNT composites. Through these techniques it is possible to extract information about functionalization properties and structural characteristics of the composites.

Table 1 Most common analytical techniques used to characterize functionalized CNTs under oxide supports

Technique	Analysis	Reports
TGA	Thermal stability and organic material decomposition	[50, 58, 60, 61]
SEM	Morphology: Clusters and degree of coating	[2, 11, 52, 53, 55, 60–63, 67]
TEM	Morphology; Analysis of impurities; Oxide support on CNT	[50, 52, 55, 60, 61, 63–65, 67]
Raman spectroscopy	Relative amount of carbon impurities and damage/ disorder; Chirality of SWCNT	[52, 58, 63, 64, 67]
Photocatalysis	Photoactivity of the composite: degradation of organic dye under UV/visible radiation	[2, 11, 50, 55, 60–64, 68]
Specific surface area	Change in the value of specific surface area of the support oxide	[58, 62, 63, 68]
XRD	Phase identification in the composite; Effect of carbon content on the crystallization of TiO_2	[2, 11, 50, 55, 58, 60–65, 68]
FT-IR	Formation of CNT/TiO_2 bonds	[52, 60, 65]
EDX	Mapping of the distribution of elements	[11, 63, 65]

References

1. Seeger, T., Redlich, Ph., Grobert, N., et al.: SiO_x-coating of carbon nanotubes at room temperature. Chem. Phys. Lett. **339**, 41–46 (2001)
2. Oh, W.C., Chen, M.L.: Synthesis and characterization of CNT/TiO_2 composites thermally derived from MWCNT and titanium (IV) n-butoxide. Bull. Korean Chem. Soc. **29**, 159–164 (2008)
3. Kroto, H.W., Heath, J.R., O'Brien, S.C., Curl, R.F., Smalley, R.E.: C_{60}: buckminsterfullerene. Nature **318**, 162–163 (1985)
4. Dresselhaus, M.S., Dresselhaus, G., Saito, R.: Physics of carbon nanotubes. Carbon **33**, 883–891 (1995)
5. Iijima, S.: Helical microtubules of graphitic carbon. Nature **354**, 56–58 (1991)
6. Ajayan, P.M., Zhou, O.Z.: Carbon nanotubes, topics appl. phys. 80. In: Dresselhaus, M.S., Dresselhaus, G., Avouris, P. (eds.) Applications of carbon nanotubes, Springer-Verlag, New York (2001)
7. Hou, P.X., Liu, C., Cheng, H.M.: Purification of carbon nanotubes. Carbon **46**, 2003–2025 (2008)
8. Hillert, M., Lange, N.: The structure of graphite filaments. Z. Kristallogr. **111**, 24–34 (1958)
9. Iijima, S., Ichibashi, T.: Single shell carbon nanotubes of one nanometer diameter. Nature **363**, 603–605 (1993)
10. Serp, P., Corrias, M., Kalck, P.: Carbon nanotubes and nanofibers in catalysis. Appl. Catal. A **253**, 337–358 (2003)
11. Chen, M.L., Zhang, F.J., Oh, W.C.: Synthesis, characterization, and photo catalytic analysis of CNT/TiO_2 composites derived from MWCNTs and titanium sources. New Carbon Mater. **24**, 159–166 (2009)
12. http://www.aerosil.com/product/aerosil/en/effects/photocatalyst

13. Saint-Gobain Glass France.: Substrate with a self-cleaning coating. Patent WO 2003/087002 (2003)
14. Funakoshi, K., Nonami, T.: Photocatalytic treatments on dental mirror surfaces using hydrolysis of titanium alkoxide. J. Coat. Technol. Res. **4**, 327–333 (2007)
15. Chung, C.J., Lin, H., Tsou, H.-K., Shi, Z.-Y., He, J.-L.: An antimicrobial TiO$_2$ coating for reducing hospital-acquired infection. J. Biomed. Mater. Res. B Appl. Biomater. **85B**(1), 220–224 (2007)
16. Veronovski, N., Rudolf, A., Smole, M.S., Kreže, T., Geršak, J.: Self-cleaning and handle properties of TiO$_2$-modified textiles. Fibers Polym. **10**, 551–556 (2009)
17. Hashimoto, K., Irie, H., Fujishima, A.: TiO$_2$ photocatalysis: a historical overview and future prospects. Jpn. J. Appl. Phys. **44**, 8269–8285 (2005)
18. Shen, G., Chen, P.C., Ryu, K., Zhou, C.: Devices and chemical sensing applications of metal oxide nanowires. J. Mater. Chem. **19**, 828–839 (2009)
19. Haggfeldt, A., Bjorksten, U., Lindquist, S.E.: Photoelectrochemical studies of colloidal TiO$_2$-films: the charge separation process studied by means of action spectra in the UV region. Sol. Energy Mater. Sol. Cells **27**, 293–304 (1992)
20. Hermass, J.M.: Heterogeneous photocatalysis: fundamentals and applications to the removal of various types of aqueous pollutants. Catal. Today **53**, 115–129 (1999)
21. Ha, H.K., Yosimoto, M., Koinuma, H., Moon, B., Ishiwara, H.: Open air plasma chemical vapor deposition of highly dielectric amorphous TiO$_2$ films. Appl. Phys. Lett. **68**, 2965–2967 (1996)
22. Bahtat, A., Bouderbala, M., Bahtat, M., Bouazaoui, M., Mugnier, J., Druetta, M.: Structural characterisation of Er^{3+} doped sol-gel TiO$_2$ planar optical waveguides. Thin Solid Films **323**, 59–62 (1998)
23. Desu, S.B.: Ultra-thin TiO$_2$ films by a novel method. Mater. Sci. Eng. B **13**, 299–303 (1992)
24. Sato, T., Taya, M.: Copper-aided photosterilization of microbial cells on TiO$_2$ film under irradiation from a white light fluorescent lamp. Biochem. Eng. J. **30**, 199–204 (2006)
25. Tamai, H., Katsu, N., Ono, K., Yasuda, H.: Simple preparation of TiO$_2$ particles dispersed activated carbons and their photo-sterilization activity. J. Mater. Sci. **37**, 3175–3180 (2002)
26. Ditta, I.B., Steele, A., Liptrot, C., Tobin, J., Tyler, H., Yates, H.M., Sheel, D.W., Foste, H.A.: Photocatalytic antimicrobial activity of thin surface films of TiO$_2$, CuO and TiO$_2$/CuO dual layers on *Escherichia coli* and bacteriophage T4. Appl. Microbiol. Biotechnol. **79**, 127–133 (2008)
27. Lei, S., Guo, G., Xiong, B., Gong, W., Mei, G.: Disruption of bacterial cells by photocatalysis of montmorillonite supported titanium dioxide. J. Wuhan Univ. Technol. Mat. Sci. Ed. **24**, 557–561 (2009)
28. Cui, H., Jiang, J., Gu, W., Sun, C., Wu, D., Yang, T., Yang, G.: Photocatalytic inactivation efficiency of anatase nano-TiO$_2$ sol on the H9N2 avian influenza virus. Photochem. Photobiol. **86**, 1135–1139 (2010)
29. Byrne, J.: Using nanotechnology to improve photocatalytic efficiencies for water treatment. http://www.azonano.com/article.aspx?ArticleID=2410. (2009)
30. Su-juan, Z., Ling-fang, Y., Dan, S., Ying-jie, W.: Effect of Cs+, Ag+, Fe^{3+} doping and Ag$^+$, Fe^{3+} co-doping contents on photocatalytic activity of TiO$_2$ films. 2011 symposium on photonics and optoelectronics (SOPO), Wuhan, China (2011)
31. Dang, TMD., Le, D.D., Chau, V.T., Dang, M.C.: Visible-light photocatalytic activity of N/SiO$_2$–TiO$_2$ thin films on glass. Adv. Nat. Sci: Nanosci. Nanotechnol. **015004**, 5 (2010)
32. Li, D., Haneda, H., Labhsetwar, N.K., Hishita, S., Ohashi, N.: Visible-light-driven photocatalysis on fluorine-doped TiO$_2$ powders by the creation of surface oxygen vacancies. Chem. Phys. Lett. **401**, 579–584 (2005)
33. Li, D., Ohashi, N., Hishita, S., Kolodiazhnyi, T., Haneda, H.: Origin of visible-light-driven photocatalysis: A comparative study on N/F-doped and N-F-codoped TiO2 powders by means of experimental characterizations and theoretical calculations. J. Solid State Chem. **178**, 3293–3302 (2005)

34. Fan, H.J., Lu, C.S., Lee, W.L.W., Chiou, M.R., Chen, C.C.: Mechanistic pathways differences between P25-TiO_2 and Pt-TiO2 mediated CV photodegradation. J. Hazard. Mater. **185**, 227–235 (2011)
35. Luo, Y., Heng, Y., Dai, X., Chen, W., Li, J.: Preparation and photocatalytic ability of highly defective carbon nanotubes. J. Solid State Chem. **182**, 2521–2525 (2009)
36. Gao, Y., Liu, H., Ma, M.: Preparation and photocatalytic behavior of TiO_2-carbon nanotube hybrid catalyst for acridine dye decomposition. React. Kinet. Catal. Lett. **90**, 11–18 (2007)
37. Hu, G., Meng, X., Feng, X., Ding, Y., Zhang, S., Yang, M.: Anatase TiO_2 nanoparticles/ carbon nanotubes nanofibers: preparation, characterization and photocatalytic properties. Mater. Sci. **42**, 7162–7170 (2007)
38. Wang, Q., Yang, D., Chen, D., Wang, Y., Jiang, Z.: Synthesis of anatase titania-carbon nanotubes nanocomposites with enhanced photocatalytic activity through a nanocoating-hydrothermal process. J. Nanopart. Res. **9**, 1087–1096 (2007)
39. Meng, L., Fu, C., Lu, Q.: Advanced technology for functionalization of carbon nanotubes. Prog. Nat. Sci. **19**, 801–810 (2009)
40. Cveticanin, J., Krkljes, A., Kacarevic-Popovic, Z., et al.: Functionalization of carbon nanotubes with silver clusters. Appl. Surf. Sci. **256**, 7048–7055 (2010)
41. Jeong, G.H.: Surface functionalization of single-walled carbon nanotubes using metal nanoparticles. Trans. Nonferr. Met. Soc. China **19**, 1009–1012 (2009)
42. Xu, G., Zhu, B., Han, Y., Bo, Z.: Covalent functionalization of multi-walled carbon nanotube surfaces by conjugated polyfluorenes. Polymer **48**, 7510–7515 (2007)
43. Zhou, Z., Wang, S., Lu, L., Zhang, Y., Zhang, Y.: Functionalization of multi-wall carbon nanotubes with silane and its reinforcement on polypropylene composites. Compos. Sci. Technol. **68**, 1727–1733 (2008)
44. Yang, F., Li, Y., Zhang, S., Tao, M., Zhao, J., Hang, C.: Functionalization of multiwalled carbon nanotubes and related polyimide/carbon nanotubes composites. Synth. Met. **160**, 1805–1808 (2010)
45. Chang, C.M., Liu, Y.L.: Functionalization of multi-walled carbon nanotubes with non-reactive polymers through an ozone-mediated process for the preparation of a wide range of high performance polymer/carbon nanotube composites. Carbon **48**, 1289–1297 (2010)
46. Singh, K.V., Pandey, R.R., Wang, X., et al.: Covalent functionalization of single walled carbon nanotubes with peptide nucleic acid: nanocomponents for molecular level electronics. Carbon **44**, 1730–1739 (2006)
47. Ye, J.S., Liu, A.L.: Functionalization of carbon nanotubes and nanoparticles with lipid. Adv. Planar Lipid Bilayers Liposomes **8**, 201–224 (2008)
48. Zhang, W., Silva, S.R.P.: Reversible functionalization of multi-walled carbon nanotubes with organic dyes. Scripta Mater. **63**, 645–648 (2010)
49. Saleh, T.A., Gupta, V.K.: Functionalization of tungsten oxide into MWCNT and its application for sunlight-induced degradation of rhodamine B. J. Colloid Interface Sci. **362**, 337–344 (2011)
50. Gao, B., Chen, G.Z., Li Puma, G.: Carbon nanotubes/titanium dioxide (CNTs/TiO2) nanocomposites prepared by conventional and novel surfactant wrapping sol–gel methods exhibiting enhanced photocatalytic activity. Appl. Catal. B **89**, 503–509 (2009)
51. Jitianu, A., Cacciaguerra, T., Benoit, R., Delpeux, S., Béguin, F., Bonnamy, S.: Synthesis and characterization of carbon nanotubes—TiO2 nanocomposites. Carbon **42**, 1147–1151 (2004)
52. Aryal, S., Kim, C.H., Kim, K.W., Khil, M.S., Kim, H.Y.: Multi-walled carbon nanotubes/ TiO_2 composite nanofiber by electrospinning. Mater. Sci. Eng. C **28**, 75–79 (2008)
53. Sampaio, M.J., Silva, C.G., Marques, R.R.N., Silva, A.M.T., Faria, J.L.: Carbon nanotube–TiO_2 thin films for photocatalytic applications. Catal. Today **161**, 91–96 (2001)
54. Chen, L., Zhang, B.L., Qu, M.Z., Yu, Z.L.: Preparation and characterization of CNTs–TiO_2 composites. Powder Technol. **154**, 70–72 (2005)
55. Jiang, G., Zheng, X., Wang, Y., Li, T., Sun, X.: Photo-degradation of methylene blue by multi-walled carbon nanotubes/TiO_2 composites. Powder Technol. **207**, 465–469 (2011)

56. Belin, T., Epron, F.: Characterization methods of carbon nanotubes: a review. Mater. Sci. Eng. B **119**, 105–118 (2005)
57. Xu, J.M., Zhang, X.B., Li, Y., Tao, X.Y., Chen, F., Li, T., Bao, Y., Geise, H.J.: Preparation of Mg$_{1-x}$Fe$_x$MoO$_4$ catalyst and its application to grow MWNTs with high efficiency. Diam. Relat. Mater. **13**, 1807–1811 (2004)
58. Yu, Y., Yu, J.C., Yu, J.G., et al.: Enhancement of photocatalytic activity of mesoporous TiO$_2$ by using carbon nanotubes. Appl. Catal. A **289**, 186–196 (2005)
59. Hu, C., Duo, S., Zhang, R., Li, M., Xiang, J., Li, W.: Nanocrystalline anatase TiO$_2$ prepared via a facile low temperature route. Mater. Lett. **64**, 2040–2042 (2010)
60. Wang, S., Zhou, S.: Photodegradation of methyl orange by photocatalyst of CNTs/P-TiO$_2$ under UV and visible-light irradiation. J. Hazard. Mater. **185**, 77–85 (2011)
61. Wang, W., Serp, P., Kalck, P., et al.: Preparation and characterization of nanostructured MWCNT-TiO$_2$ composite materials for photocatalytic water treatment applications. Mater. Res. Bull. **43**, 958–967 (2008)
62. Bouazza, N., Ouzzine, M., Lillo-Ródenas, M.A., Eder, D., Linares-Solano.: TiO2 nanotubes and CNT–TiO$_2$ hybrid materials for the photocatalytic oxidation of propene at low concentration. Appl. Catal. B **92**, 377–383 (2009)
63. Zhang, K., Zhang, F.J., Chen, M.L., Oh, W.C.: Comparison of catalytic activities for photocatalytic and sonocatalytic degradation of methylene blue in present of anatase TiO$_2$–CNT catalysts. Ultrason. Sonochem. **18**, 765–772 (2011)
64. Zevellos-Márquez, A.M.O., Brasil, M.J.S.P., Likawa, F., et al.: Effect of TiO$_2$ nanoparticles on the thermal properties of decorated multiwall carbon nanotubes: a Raman investigation. J. App. Phys. **108**, 083501 (2010)
65. Zhou, W., Pan, K., Qu, Y., et al.: Photodegradation of organic contamination in wastewaters by bonding TiO$_2$/single-walled carbon nanotube composites with enhanced photocatalytic activity. Chemosphere **81**, 555–561 (2010)
66. Hu, C., et al.: Synthesis of carbon nanotube/anatase tiatania comby a combination of sol-gel and self-assembly at low temperature. J. Sol. Sta. Chem. **184**, 1286–1292 (2011)
67. Chen, L.C., Ho, Y.C., Guo, W.S., Huang, C.M., Pan, T.C.: Enhanced visible light-induced photoelectrocatalytic degradation of phenol by carbon nanotube-doped TiO$_2$ electrodes. Electrochim. Acta **54**, 3884–3891 (2009)
68. Orlanducci, S., Sessa, V., Terranova, M.L., Battiston, G.A., Battiston, S., Gerbasi, R.: Nanocrystalline TiO$_2$ on single walled carbon nanotube arrays: towards the assembly of organized C/TiO2 nanosystems. Carbon **44**, 2839–2843 (2006)

Synthesis of Vertically Aligned Carbon Nanotubes by CVD Technique: A Review

A. G. Osorio, A. S. Takimi and C. P. Bergmann

Abstract Vertically aligned carbon nanotubes are bundles of carbon nanotubes that grow perpendicular to a substrate, assembling a bamboo forest. They are dense and orderly arranged, and can be very long. The unique properties of carbon nanotubes along with the ability to produce yarns from these nanotubes forests enable vertically aligned nanotubes to be used in various fields. The most used method to synthesize these nanotubes is the CVD technique. According to the parameters used, carbon nanotubes forests will grow longer or shorter, more or less aligned. This review explores the CVD method used to obtain vertically aligned nanotubes as well as the process parameters that determine the morphology, orientation and growth size of nanotubes and the applications of these materials.

1 Background on Carbon Nanotubes

Carbon nanotubes (CNTs) are nanometric tubes of graphitic carbon with outstanding properties. They are among the stiffest and strongest fibres known, and have remarkable structural, mechanical, electrical, and thermal properties [1–3]. For these reasons they have attracted huge academic and industrial interest and hundreds of papers on nanotubes are being published every year.

Discovered in 1990 by Sumio Iijima [4], such material brought a vast range of potential applications which raised several research possibilities. There are two main types of CNTs: Single and Multi Wall CNT, named SWCNTs and MWCNTs, respectively. Their number of walls and structure, nonetheless, provide different properties.

A. G. Osorio (✉) · A. S. Takimi · C. P. Bergmann
Universidade Federal do Rio Grande do Sul, Porto Alegre, Brazil
e-mail: osorio.alice@gmail.com

C. Avellaneda (ed.), *NanoCarbon 2011*, Carbon Nanostructures,
DOI: 10.1007/978-3-642-31960-0_7, © Springer-Verlag Berlin Heidelberg 2013

There are three main techniques used to produce CNTs, they are laser ablation, arc discharge and chemical vapor deposition (CVD). Laser ablation and arc discharge's main advantage relies on the fact that it does not require a substrate to synthesize nanotubes and CNTs' purity is significantly higher when compared to CVD technique. However, their cost of production is higher due to a sophisticate apparatus needed. Nonetheless, this text reviews the synthesis of vertically aligned CNTs by CVD technique.

2 Synthesis of CNTs by Chemical Vapor Deposition and Their Variables

The CVD method allows synthesis of CNTs at lower temperatures compared to arc discharge and laser ablation, thus it is the most suitable technique for applications such as electronics. CVD usually allows synthesis of longer CNTs (up to centimeters) but with more defects compared to higher temperature processes, although a post-annealing process at temperatures close to 2000 °C may recover the crystalline structure of CNTs produced at higher temperature [5, 6].

The CVD chemical process is based upon the decomposition of carbon-containing gases on catalysts. Usually, 3d valence transition metal nanoparticles, such as Fe, Co an d Ni or alloys, are used to catalyze CNT growth. In this model, hydrocarbons adsorbed on the metal nanoparticle are catalytically decomposed resulting in atomic carbon dissolving into the liquid catalyst particle, and when a supersaturated state is reached, carbon precipitates in a tubular, crystalline form [7].

According to Hart et al. [8] CNT growth by CVD technique involves surface and/or bulk diffusion of carbon at a metal catalyst particle. CNT–CNT or CNT-substrate interactions in addition to the arrangement and activity of the catalytic sites determine if CNTs grow in an isolated, tangled, or aligned configuration. For example, isolated SWNTs can be grown to several millimeters or centimeters when suspended during growth by a gas flow; however, the density of catalytic sites must be very low to prevent entanglement between CNTs. At a high catalyst density and CNT growth rate, the CNTs grow self-oriented perpendicular to the substrate surface due to initial crowding and continue to grow upward in this direction [8]. The CVD process uses an energy source (e.g., a furnace with resistive or inductive heater, hot filament, plasma, etc.) to decompose the precursor gases and to heat the sample. In thermal CVD, the heat of the furnace is responsible for the decomposition of the precursor gases; in addition it affects the catalytic function of the substrate. Other CVD systems such as plasma-enhanced CVD (PECVD) uses plasma to aid the gas decomposition, which often allows CNTs grow at lower temperatures [9]. Hot-filament CVD technique uses the heat of the filament to support the catalysis, whereas the gas decomposition requires a second heat source before being purged into the tube where CNT growth will occur. The growth temperature typically varies between 700 °C and 900 °C as high temperature is necessary for the catalysis to occur. Recently, encouraging results have been obtained to reduce this temperature to lower values (400–550 °C) [10, 11].

More recently, publications about highly oriented, free-standing sheets of CNTs were published [12–17] and urged a whole new front of research and applications for CNTs. This synthesis process consists of a solid state process that presents potential to be scalable for continuous production. Zhang et al. [12] confirmed that meter-long sheets, up to 5 cm wide could be achieved. Such growth of aligned CNTs is based on the idea that a high density of catalyst will lead the nanotubes to grow vertically. This special arrangement of nanotube arrays is called forest because they look like a bamboo forest.

Typically, the growth of aligned CNTs by CVD technique can be achieved using an unsaturated hydrocarbon gas as the source of carbon (e.g. C_2H_2, C_2H_4) and a pre-deposited catalyst film.

Hart et al. [18] developed a simplified way of synthesizing forests of CNTs, where the reactions of synthesis take place locally on a suspended substrate that is heated resistively. Some researchers called this technique the "hot filament assisted CVD". This method, however, presents the need of a second tank in order to heat the gases previous to the synthesis. A film of up to 3 mm thick of aligned CNT forest was obtained in just 15 min, using this method.

Lately the addition of small amount of water vapor during the CVD synthesis was reported as a path to increase the growth rate, the length and the purity of CNT forests [16, 17, 19, 20]. Hata et al. [20] concluded that water acts in promoting and preserving the catalytic activity which enhances the growth of tubes. Liu et al. [17] evidenced that water play a role in the purity and surface defects of synthesized CNTs.

Doping of CNTs during CVD synthesis has also attracted researchers' interest and significant results are being recently published concerning this topic. By doping CNTs it is possible to changes their chemical and physical characteristics, for electrical applications, e.g., it is crucial to control the conduction type (n or p). Up to date, the most studied chemical elements for the doping of CNTs is boron (B) and nitrogen (N) [21, 22], they are the neighbor of carbon at the periodic table and would theoretically provide p- and n- doping, respectively. Lee et al. [22] studied the use of B-doping as an inhibitor for CNTs oxidation and depicted that boron can act as an inhibitor, although the structure of CNTs is changed with the addition of such element. Phosphorus (P) is also an element quite used for doping of nanotubes [23–25].

Apart from the high efficiency and relative low cost of production, vertically aligned CNTs can also be spun into yarns [13, 19, 26–28]. The CNTs can be pulled into low density yarns at speeds of up to meters per second. Huynh et al. [13] showed that spinnability of CNT forests is extremely sensitive to synthesis parameters, such as catalyst, substrate, temperature, gas flow rates, reaction time etc. The effect of each of these parameters will be discussed later on.

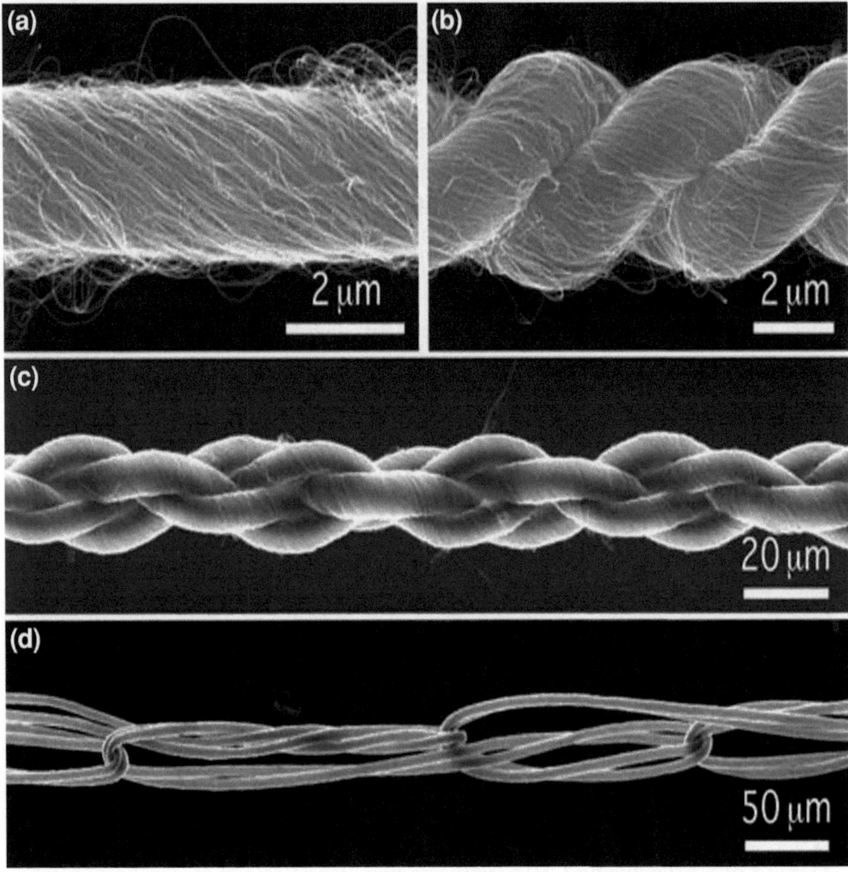

Fig. 1 SEM images of twisted and knitted CNTs [26]

Once discovered that CNT forests could be spun into ribbons or yarns, an interest on applying such yarn has grown into researchers. Zhang et al. [26] introduced twist during the spinning of nanotube forests and registered yarn strength greater than 460 MPa. CNT yarns were also knitted and knotted introducing a potential use in textile industry (Fig. 1).

2.1 Study of the Parameters of Vertically Aligned CNT Forests

The alignment of CNTs will depend on several factors/parameters; among them are the roughness of the substrate, the catalyst used, parameters of the synthesis and so on. The most significant issues will be discussed below.

2.1.1 Influence of the Substrate

The use of a minimal roughness and homogeneity of the substrate showed as the most qualified surface to grow aligned CNTs. The material composition of the substrate does not play any large influence on CNTs' growth because a buffer layer is usually deposited over the substrate, which isolates it from the catalyst. Nonetheless, a high uniformity of the surface is crucial to nucleate and grow vertical CNTs.

Several researchers evidenced the need of deposition of an insulating layer on the top of the substrate previous to the deposition of catalyst in order to avoid any diffusion between the catalyst and the substrate. In most cases when the substrate used is metallic, an oxide layer is thermally grown. The size needed to guarantee the insulation of the substrate is widely discussed but no agreement is reached. De los Arcos et al. [29] studied the influence of different buffer layers and concluded that Al_2O_3, TiN and TiO_2 can be used to grow aligned CNTs whereas Al layers cannot. On the other hand, Choi et al. [30] depicted results that indicate that Al buffer is very efficient to grow aligned CNTs.

In general, according to literature, one can say that the most efficient buffer layer is the thermally grown or deposited oxide formed from the metal on request [31, 32]. Huynh et al. [13] tested a variety of SiO2 thickness on a silicon substrate ranging from a native oxide to 900 nm and depicted that any thickness higher that 25 nm resulted on long CNTs' growth, whereas the native oxide presented rather poorer length of CNTs. In addition, the author also concluded that the oxide thickness does not seem to play a role on the length of nanotubes; however, the spinnability of such tubes is widely dependent on the thickness of the oxide layer. The authors depicted that a layer of 25–50 nm is the ideal size to obtain long and spinnable CNTs.

2.1.2 Influence of the Catalyst

The catalyst plays an important role on the growth of aligned CNTs, not only because it is the starting point of nucleation but also because it can control the vertical growth by its saturation, homogeneity, thickness of the layer as well as the diameter of nanotubes. But what is the driving force that leads the growth of CNTs from a metal catalyst?

Previous papers explained that the decomposition of the carbon source is followed by the formation of metal carbides. Once the surface of the particle is completely carburized, this carburized layer acts as a barrier for further carbon diffusion. This gives rise to the precipitation of carbon at the metal carbide surface, driving the growth of carbon structures [33, 34].

It is known that the catalytic activity of a metal is highly dependent of its electronic structure. According to Esconjauregui et al. [34], the reactivity of a metal particle is determined by the number of electron vacancies in the d-orbital of the metal. As the reactivity exceeds a threshold value, carbon atoms lose their mobility in the solid solution and decompose into metal carbides. Thus, elements with no d-vacancies, as Cu and Zn, are considered inert to carbon; whereas elements with many d-vacancies, as Ti and V, show very strong interaction with carbon which will result in the formation of stable carbides. At last, elements with few d-vacancies exhibit moderate interaction with carbon, forming metastable carbides that will eventually block the diffusion of further carbon and starts to precipitate it in a graphite structure. Among this group of selected metals, one can mention Ni, Co and Fe as the most used metal particles to nucleate and grow nanotubes, highlighting Fe as the most used [12–14]. Esconjauregui et al. [34] proposed, therefore, that Ni, Co and Fe are the best catalyst to nucleate CNTs due to their electronic structure that presents few d-vacancies.

Another issue that raised several studies concerning the catalyst for CNTs' growth was the catalyst particle size/thickness of the layer. Does it influence CNTs diameter and or length? Substrates were submitted to studies and a deep evaluation of it was performed by Moodley et al. [31] where the authors tested iron particles and concluded that these particles are involved in a series of re-arrangements previous to the initial CNTs' growth. The diameter of the nanotubes, therefore, is not related to the initial catalyst particle size. On the other hand, other researchers showed that a catalyst layer as thick as 1 nm or lower does not present large amount of CNTs as well as vertically aligned nanotubes [13]. This fact can be attributed to the insufficient amount of catalyst to cover the entire substrate; when the catalyst is deposited, it tends to agglomerate into nanoparticle islands and holes among these metal islands may be formed and prevent the carbon nucleation. Thus, a less dense CNT forest is obtained when short catalyst layer is used due to the lack of continuity of the catalyst layer, because the alignment of the nanotubes is related to the steric hindrance between closely spaced catalyst particles. Yet, some researchers depicted the influence and possible control of nanotubes diameter by the control of catalyst particles [35].

2.1.3 Influence of the Temperature and Reaction Time of CVD Synthesis

The temperature is mainly dependent on the reactor being used, such as its design, diameter of the quartz tube, whether it is a one or three zones furnace, etc. Reviewing the literature one will notice that CNTs can grow from temperatures as low as 400 °C to temperatures as high as 900 °C. The average reaction temperature found on literature is in the range of 650–850 °C. Moreover, the reaction temperature does not seem to have any relevant relation with the growth of CNT forests.

The influence of the reaction time, however, is rather significant on the growth of aligned CNTs. In general, longer reaction times will result in longer CNTs [13, 32]. A maximum growth time, however, was observed by some researcher [32]. The hypothesis that explains such behavior says that above this time the catalyst loses its activity, i.e. the metal particles are not active anymore, and it may happen due to the encapsulation of the catalyst particles that would deactivate them.

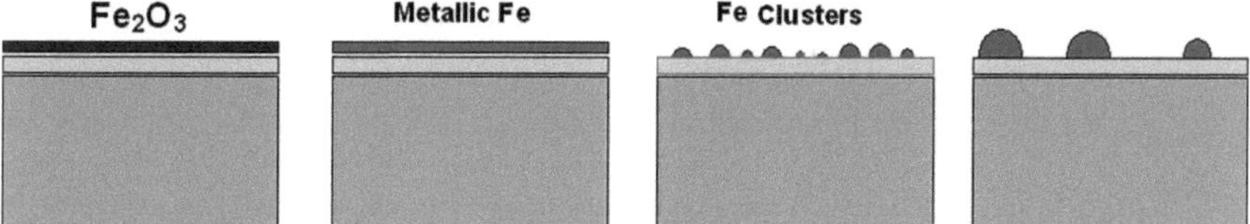

Fig. 2 Illustration of the coarsening of the catalyst during the reduction

2.1.4 Influence of Gas Flow Rates

There are different gases used as precursor during the synthesis of CNTs, among them one can cite ethylene, acetylene, methane and hexane. Apart from the precursor gas, a carrier gas such as argon or helium is used and, in some cases, a reducing gas is needed in order to reduce the oxide metal catalyst previous to the carbon diffusion and precipitation.

The flow rate is seen in a vast range, depending whether the carbon source gas used it is more or less reactive, i.e. if the gas used is acetylene, the amount of acetylene used is about 5 % of acetylene [13, 26], which is already sufficient to react. In other cases, this value may be much higher, according to how reactive the gas is.

There are also studied concerning the use of the reducing gas; the most used is hydrogen because of its ability to enable catalyst reduction and also reduce gas-phase pyrolysis of hydrocarbons. In addition, hydrogen can cause desorption of amorphous fragments from the surface of the catalysts. Nessim GD et al. [36] studied the influence of timing of hydrogen exposure to the catalyst film on the growth of CNTs' forests and concluded that there is an optimum time where a better growth of CNT forests is obtained. The author suggests that once hydrogen is introduced, the iron oxide will be reduced and iron clusters will form. These clusters are formed by film agglomeration after reduction of the iron oxide; initially, they are small and closely spaced. Continued exposure to hydrogen leads the clusters to coarsen (Fig. 2). This process is driven by surface energy minimization. During the coarsening process, small clusters tend to shrink and disappear with their mass being redistributed to larger clusters. This process therefore results in an increase in the average cluster size as well as an increase in the average cluster spacing, which directly influence on the growth and alignment of CNTs [36].

3 Applications

3.1 Carbon Nanotube Fibers and Yarns

To make use of the extraordinary properties of CNTs, is often necessary arrange the nanotubes in a macroscopic yarn. Spinning carbon nanotube fibers directly from carbon nanotubes forests (solid-state spinning) is one of the several ways to fabricate high-quality fibers [37–39]. In this method, a bundle of CNTs are drawn from a carbon nanotube forest, and due to the entanglement and interaction forces between the adjacent nanotubes (mainly van der Waals force) [40], more bundles of CNTs are draw in a continuous way, forming a continuous fiber.

The properties of a CNT fiber are dictated by the strength of the constituent bundles (or ropes) and their connectivity. Even when single nanotube is perfectly aligned into a single bundle, the failure of this assembly can occur via the slippage of the constitutive CNTs at a stress lower than the failure stress of the single nanotube. Fibers made by solid-state spinning can achieve modulus up to 330 GPa and 1,91 GPa of strength [41].

This fiber can be spun to make a strand, and subsequently used in textiles, composites, electrical cables and many more.

3.2 Supercapacitors

Supercapacitors are electrochemical energy storage devices that combine the high energy-storage-capability of conventional batteries with the high power-delivery-capability of conventional capacitors. Supercapacitors have been developed to provide power pulses for a wide range of applications including electric transportation technology (hybrid electric vehicles), electric utility industry (emergency backup power and grid system stability improvement), consumer electronics (laptops, cell phones, camcorder), medical electronics (portable defibrillators, drug delivery units, and neurological stimulators), and several military/defense devices.

CNTs posses several properties that make suitable for developing high-performance supercapacitors, like high electrical conductivity, high specific surface area, high charge transport capability, high mesoporosity, and high electrolyte accessibility [42]. Vertically aligned carbon nanotubes are advantageous over their randomly entangled CNTs in this application.

Second, aligned CNTs have enhanced effective surface area. Surface area of randomly entangled CNTs is determined by the open space between entangled fibrils. Similarly, surface area in vertically aligned CNTs is determined by the open space between aligned tubes, and is in the range of tens of nanometer7. This provides a well-defined surface area for each of the constituent tubes to be accessible to the electrolyte ions, resulting in enhanced effective surface area for aligned CNTs.

3.3 Chemical Sensors

Gas adsorption capability, large specific surface area, high sensitivity, chemical interactions, lower operating temperature and wider variety of gas detection capabilities compared to common sensors make CNTs very attractive as gas sensing nanomaterials [42, 43]. Possible chemical sensor applications of nonmetallic nanotubes are interesting, because nanotube electronic transport and thermopower (voltages between junctions caused by interjunction temperature differences) are very sensitive to substances that affect the amount of injected charge (chemiresistive sensing) [41, 43].

Aligned carbon nanotube arrays, in either a patterned or nonpatterned form, allow the development of novel sensors and/or sensor chips without the need for direct manipulation of individual nanotubes since the constituent nanotubes can be collectively addressed through a common substrate/electrode. The aligned nanotube structure further provides a large well-defined surface area and the capacity for modifying the carbon nanotube surface with various transduction materials to effectively enhance the sensitivity and to broaden the scope of analytes to be detected [42, 43]. In addition, vertical alignment of the CNT nanoeletrodes is preferred over other schemes because the open end of a CNT is expected to show faster electron transfer rate (ETR) [44].

Chemical sensors based in CNTs show enhanced detection level for a large number of chemical gases and substances, like NH_3, Ethanol, SO_2, CH_4, CO, and many other gases/agents. Sinha et al. [45] provide a comprehensive overview about the use of carbon nanotube-bases sensor over a wide variety of substances.

3.4 Biomimetic Dry Adhesive

Dry adhesion mechanism in gecko feet has attracted much attention since it provides strong, yet reversible attachment against surfaces of varying roughness and orientation. Such unusual adhesion capability is attributed to arrays of millions of fine microscopic foot hairs (setae), splitting into hundreds of smaller, nanoscale ends (spatulae), which form intimate contact to various surfaces by van der Waals forces with strong adhesion ($\sim$10 N/cm^2) [46]. Since the discovery of the major role of van der Waals forces in gecko adhesion, extensive studies have been made to develop a new gecko-inspired artificial dry adhesive by mimicking gecko foot hairs, instead of using a viscoelastic polymer-based adhesive.

Yurdumakan et al. measured adhesion of the vertically aligned carbon nanotubes (grown by CVD over SiO_2 substrate and transferred to a PMMA film) at the nanometer level by using a scanning probe microscope (SPM). They found that the adhesion force is 200 times higher than that of gecko foot hairs [47]. Ge et al. make a 1-cm^2 area flexible tape (gecko tape) with microfabricated multiwalled carbon nanotube forests transferred to a flexible tape, and found a shear adhesion force of 36 N, which is four times higher than the natural gecko foot-hairs [48].

3.5 *Lotus Effect*

The Lotus Effect refers to an extremely high water repellence (called superhydrophobicity), when the contact angle of a water droplet is higher than 150° [49]. These properties are desirable for many industrial and biological applications such as self-cleaning paints for buildings and houses, antifouling paints for boats and ships, stain resistant textiles, and many more.

To make it possible, the surface has to possess a combination of microstructure and nanostructure on their surface which minimizes the contact area with anything that came into contact with the surface. In the case of Lotus leaf, microstructural epidermal cells create a rough surface, and these epidermal cells are covered with nanometric wax crystals. The wax crystals provide a water-repellent layer, which is enhanced by the surface roughness according to the Wenzel and Cassie-Baxter models. As results, water droplets on the surface tend to assume a near-spherical format to minimize the contact area with the surface [48].

As-synthesized vertically aligned carbon nanotubes are almost superhydrophobic (contact angle 142 ± 6°) but this level of hydrophobicity is not stable [50]. To turn a CNT forest in a superhydrophobic surface, carbon nanotubes must be coated with a low surface energy material like PTFE or a silane [51, 52].

References

1. Meyyappan, M.: Carbon nanotubes: science and applications, pp 2–21. NASA AR Center. Moffett Field, CA. CRC Press LLC (2005)
2. Gogotsi, Y.: Carbon Nanomaterials, pp 41–70. Drexel University, Philadelphia, Pennsylvania, USA. Taylor and Francis Group, LLC (2006)
3. Thostenson, E.T., Ren, Z., Chou, T.-W.: Advances in the science and technology of carbon nanotubes and their composites: a review. Comp. Sci. Technol. **61**, 1899–1912 (2001)
4. Iijima, S.: Helical microtubules of graphitic carbon. Nature **354**, 56 (1991)
5. Pinault, M., Mayne-L'Hermite, M., Reynaud, C., Beyssac, O., Rouzaud, J.N., Clinard, C.: Carbon nanotubes produced by aerosol pyrolysis: growth mechanisms and post-annealing effects. Diam. Relat. Mater. **13**, 1266–1269 (2004)
6. Huang, W., Wang, Y., Luo, G., Wei, F.: 9 9.9 % purity multi-walled carbon nanotubes by vacuum high-temperature annealing. Carbon **41**, 2585–2590 (2003)
7. Marulanda, J.M.: Carbon nanotubes. In-Tech, India (2010)
8. Hart, A.J., Slocum, A.H.: Rapid growth and flow-mediated nucleation of millimeter-scale aligned carbon nanotube structures from a thin-film catalyst. J. Phys. Chem. B **110**, 8250–8257 (2006)
9. Choi, S.I., Nam, J.S., Lee, C.M., Choi, S.S., Kim, J.I., Park, J.M., Hong, S.H.: High purity synthesis of carbon nanotubes by methane decomposition using an arc-jet plasma. Curr. Appl. Phys. **6**, 224–229 (2006)
10. Jayatissa, A.H., Guo, K.: Synthesis of carbon nanotubes at low temperature by filament assisted atmospheric CVD and their field emission characteristics. Vacuum **83**, 853–856 (2009)
11. Chen, K.-C.: Low-temperature CVD growth of carbon nanotubes for field emission application. Diam. Relat. Mater. **16**, 566–569 (2007)

12. Zhang, M., Fang, S., Zakhidov, A.A., Lee, S.B., Aliev, A.E., Williams, C.D., Atkinson, K.R., Baughman, R.H.: Strong, transparent, multifunctional, carbon nanotube sheets. Science **309**, 1215–1219 (2005)
13. Huynh, C.P., Hawkins, S.C.: Understanding the synthesis of directly spinnable carbon nanotube forests. Carbon **48**, 1105–1115 (2010)
14. Li, Q., Zhang, Z., DePaula, R.F., Zheng, L., Zhao, Y., Stan, L., Holesinger, T.G., Arendt, P.N., Peterson, D.E., Zhu, T.E.: Sustained growth of ultralong carbon nanotube arrays for fiber spinning. Adv. Mater. **18**, 3160–3163 (2006)
15. Zhang, Y., Li, R., Liu, H., Sun, X., Merel, P., Desilets, S.: Integration and characterization of aligned carbon nanotubes on metal/silicon substrates and effects of water. Appl. Surf. Sci. **255**, 5003–5008 (2009)
16. Hu, J.L., Yang, C.C., Huang, J.H.: Vertically-aligned carbon nanotubes prepared by water-assisted chemical vapor deposition. Diam. Relat. Mater. **17**, 2084–2088 (2008)
17. Liu, H., Zhang, Y., Li, R., Suna, X., Wang, F., Ding, Z., Mérel, P., Desilets, S.: Aligned synthesis of multi-walled carbon nanotubes with high purity by aerosol assisted chemical vapor deposition: effect of water vapor. Appl. Surf. Sci. **256**, 4692–4696 (2010)
18. Hart, A.J., Van Laake, L., Slocum, A.H.: Desktop growth of carbon-nanotube monoliths with in situ optical imaging. Small **3**(5), 772–777 (2007). doi:10.1002/smll.200600716
19. Zhang, Q., Wang, D.-G., Huang, J.-Q., Zhou, W.-P., Luo, G.-H., Qian, W.-Z., Wei, F.: Dry spinning yarns from vertically aligned carbon nanotubes arrays produced by an improved floating catalyst chemical vapor deposition method. Carbon **48**, 2855–2861 (2010)
20. Hata, K., Futaba, D.N., Mizuno, K., Namai, T., Yumura, M., Iijima, S.: Water-assisted highly efficient synthesis of impurity-free single-walled carbon nanotubes. Science **30**, 1362–1364 (2004)
21. Suzuki, S., Hibino, H.: Characterization of doped single-wall carbon nanotubes by Raman spectroscopy. Carbon **49**, 2264–2272 (2011)
22. Lee, Y.-J., Kim, H.-H., Hatori, H.: Effects of substitutional B on oxidation of carbon nanotubes in air and oxygen plasma. Carbon **42**, 1053–1056 (2004)
23. Cruz-Silva, E., Cullen, D.A., Gu, L., Romo-Herrera, J.M., Muñoz-Sandoval, E., López-Urías, F., Sumpter, B.G., Meunier, V., Charlier, J.C., Smith, D.J., Terrones, H., Terrones, M.: Heterodoped nanotubes: theory, synthesis, and characterization of phosphorus-nitrogen doped multiwalled carbon nanotubes. ASC Nano **2**, 441–448 (2008)
24. Lee, Y.-J., Radovic, L.R.: Oxidation inhibition effects of phosphorus and boron in different carbon fabrics. Carbon **41**, 1987–1997 (2003)
25. Maciel, I.O., Campos-Delgado, J., Cruz-Silva, E., Pimenta, M.A., Sumpter, B.G., Meunier, V., Lopez-Urias, F., Munoz-Sandoval, E., Terrones, H., Terrones, M., Jorio, A.: Synthesis, electronic structure, and Raman scattering of phosphorus-doped single-wall carbon nanotubes. Nano Lett. **9**(6), 2267–2292 (2009)
26. Zhang, M., Atkinson, K.R., Baughman, R.H.: Multifunctional carbon nanotube yarns by downsizing an ancient technology. Science **306**, 1358–1361 (2004)
27. Tran, C.D., Humphries, W., Smith, S.M., Huynh, C., Lucas, S.: Improving the tensile strength of carbon nanotube spun yarns using a modified spinning process. Carbon **47**, 2662–2670 (2009)
28. Kim, J.-H., Jang, H.-S., Lee, K.H., Overzet, L.J., Lee, G.S.: Tuning of Fe catalysts for growth of spin-capable carbon nanotubes. Carbon **48**, 538–547 (2010)
29. De los Arcos, T., Garnier, M.G., Oelhafen, P., Mathys, D., Seo, J.W., Domingo, C., Garcia-Ramos, J.V., Sanchez-Cortes, S.: Strong influence of buffer layer type on carbon nanotube characteristics. Carbon **42**, 187–190 (2004)
30. Choi, B.H., Yoo, H., Kim, Y.B., Lee, J.H.: Effects of Al buffer layer on growth of highly vertically aligned carbon nanotube forests for in situ yarning. Microelectron. Eng. **87**, 1500–1505 (2010)
31. Moodley, P., Loosc, J., Niemantsverdriet, J.W., Thüne, P.C.: Is there a correlation between catalyst particle size and CNT diameter? Carbon **47**, 2002–2013 (2009)

32. Liu, K., Sun, Y., Chen, L., Feng, C., Feng, X., Jiang, K., Zhao, Y., Fan, S.: Controlled growth of super-aligned carbon nanotube arrays for spinning continuous unidirectional sheets with tunable physical properties. Nano Lett. **8**(2), 700–705 (2008)
33. Bonnet, F., Ropital, F., Berthier, Y., Marcus, P.: Filamentous carbon formation caused by catalytic metal particles from iron oxide. Mater. Corros. **54**(11), 870–880 (2003)
34. Esconjauregui, S., Whelan, C.M., Maex, K.: The reasons why metals catalyze the nucleation and growth of carbon nanotubes and other carbon morphologies. Carbon **47**, 659–669 (2009)
35. Li, Y., Kim, W., Zhang, Y., Rolandi, M., Wang, D., Dai, H.: Growth of single-walled carbon nanotubes from discrete catalytic nanoparticles of various sizes. J. Phys. Chem. B **105**, 11424–11431 (2001)
36. Nessim, G.D., Hart, J.A., Kim, J.S., Acquaviva, D., Oh, J., Morgan, C.D., Seita, M., Leib, J.S., Thompson, C.V.: Tuning of vertically-aligned carbon nanotube diameter and areal density through catalyst pre-treatment. Nano Lett. **8**(11), 3587–3593 (2008)
37. Lepró, X., Lima, M.D., Baughman, R.H.: Spinnable carbon nanotube forests grown on thin, flexible metallic substrates. Carbon **48**, 3621–3627 (2010)
38. Li, Y.-L., Kinloch, I.A., Windle, A.H.: Direct spinning of carbon nanotube fibers from chemical vapor deposition synthesis. Science **304**(5668), 276–278 (2004)
39. Stano, K.L., Koziol, K., Pick, M., Motta, M.S., Moisala, A., Vilatela, J.J., Frasier, S., Windle, A.H.: Direct spinning of carbon nanotube fibres from liquid feedstock. Int. J. Mater. Form. **1**, 59–62 (2008)
40. Gilvaei, A.F., Hirahara, K., Nakayama, Y.: In situ study of the carbon nanotube yarn drawing process. Carbon **49**, 4928–4935 (2011)
41. Behabtu, N., Green, M.J., Pasquali, M.: Carbon nanotube-based neat fibers. Nanotoday **3**(5–6), 24–34 (2008)
42. Baughman, R.H., Zakhidov, A.A., de Heer, W.A.: Carbon Nanotubes—the route toward applications. Science **297**(5582), 787–792 (2002)
43. Vairavapandian, D., Vichchulada, P., Lay, M.D.: Preparation and modification of carbon nanotubes: review of recent advances and applications in catalyst and sensing. Anal. Chim. Acta **626**(2), 119–129 (2008)
44. Wisitsoraat, A., Karuwan, C., Wong-ek, K., Phokharatkul, D., Sritongkham, P., Tuantranont, A.: High sensitivity electrochemical cholesterol sensor utilizing a vertically aligned carbon nanotube electrode with electropolimerized enzyme immobilization. Sensor **9**(11), 8658–8668 (2009)
45. Sinha, N., Ma, J., Yeow, J.T.W.: Carbon nanotube-bases sensors. J. Nanosci. Nanotechnol. **6**(3), 573–590 (2006)
46. Jeong, H.E., Suh, K.Y.: Nanohairs and nanotubes: efficient structural elements for gecko-inspired artificial dry adhesives. Nano Today **4**(4), 335–346 (2009)
47. Yurdumakan, B., Raravikar, N.R., Ajayan, P.M., Dhinojwala, A.: Synthetic gecko foot-hairs from multiwalled carbon nanotubes. Chem. Commun. **30**, 3799–3801 (2005)
48. Ge, L., Sethl, S., Ci, L., Ajayan, P.M., Dhinjwala, A.: Carbon nanotube-based synthetic gecko tapes. PNAS **26**(104), 10792–10795 (2007)
49. Li, X.-M., Reinhoudt, D., Crego-Calama, M.: What do we need for a superhydrophobic surface? A review on the recent progress in the preparation of superhydrophobic surfaces. Chem. Soc. Rev. **8**(36), 1350–1368 (2007)
50. Ramos, S.C., Vasconcelos, G., Antunes, E.F., Logo, A.O., Trava-Airoldi, V.J., Corat, E.J.: Wettability control on vertically-aligned multi-walled carbon nanotube surfaces with oxygen pulsed DC plasma and CO2 laser treatments. Diam. Relat. Mater. **19**(7–9), 752–755 (2009)
51. Lau, K.K.S., Bico, J., Teo, K.B.K., Chhowalla, M., Amaratunga, G.A.J., Milne, W.I., McKinley, G.H., Gleason, K.K.: Superhydrophobic carbon nanotube forests. Nano Lett. **3**(12), 1701–1705 (2003)
52. Bu, I.Y.Y., Oei, S.P.: Hydrophobic vertically aligned carbon nanotubes on corning glass for self cleaning applications. Appl. Surf. Sci. **256**(22), 6699–6704 (2010)

Thermoset Three-Component Composite Systems Using Carbon Nanotubes

L. V. da Silva, S. C. Amico, S. H. Pezzin, L. A. F. Coelho
and C. M. Becker

Abstract In this chapter, a brief review on three-component composites comprised of a (micro) fibrous reinforcement and a thermoset polymer resin filled with nanoparticles is presented. For this type of composite, carbon nanotubes (CNT) and carbon nanofibers are more commonly used and the focus usually lies on resin-dominated properties, such as interlaminar shear strength, and interlaminar fracture toughness. Many three-component systems comprised of fiber/epoxy/CNT have been produced using resin transfer molding (RTM) or VARTM. However, there are major difficulties associated with the impregnation of a dry fibrous reinforcement using a highly viscous suspension of resin/nanofiller, especially for high content of nanofillers or highly packed fibrous systems. In such harsh circumstances, an alternative and recent approach to enable processing comprises the production/use of three-component prepregs containing nanofillers, although they are usually associated with high cost. The presented case study focused on an alternative route to produce glass-fiber composites with high content of CNT via

L. V. da Silva (✉) · S. C. Amico · C. M. Becker
Department of Materials Engineering, Universidade Federal do Rio Grande do Sul,
Porto Alegre/RS, Brazil
e-mail: laisvasc@gmail.com

S. C. Amico
e-mail: amico@ufrgs.br

C. M. Becker
e-mail: crismbecker@yahoo.com.br

S. H. Pezzin · L. A. F. Coelho
Centro de Ciências Tecnológicas, Universidade do Estado de Santa Catarina, Joinville/SC,
Brazil
e-mail: dqm2shp@joinville.udesc.br

L. A. F. Coelho
e-mail: dma2lafc@joinville.udesc.br

C. Avellaneda (ed.), *NanoCarbon 2011*, Carbon Nanostructures,
DOI: 10.1007/978-3-642-31960-0_8, © Springer-Verlag Berlin Heidelberg 2013

RTM. A practical, low-cost and effective methodology for the direct deposition of an acetone/CNT/epoxy suspension on glass-fiber cloths was developed, achieving up to 4.15 % wt. in overall CNT content in the composite. The mechanical properties of the composites produced with non-functionalized CNT increased, in general, up to 10 % compared to the reference epoxy/glass-fiber composite. However, the high CNT content obtained was of uttermost importance for the development of electromagnetic characteristics on the material, absorbing much of the radiation in the microwave frequency range. The reflectivity properties reached a maximum of approximately -14dB (c.a. 95 % of electromagnetic absorption) and this excellent performance was obtained using a comparatively low cost (glass fiber) and thin (»2.2 mm) polymer composite material. Thus, the developed composites showed great potential to be used as microwave-absorption materials, replacing conventional ones employed for this aim. With further improvement in the manufacturing process, these materials could become of interest as high performance composites in a wide range of engineering applications, from telecommunications to aerospace.

1 Introduction

In this chapter, a brief review on three-component composites comprised of a (micro) fibrous reinforcement and a thermoset polymer resin filled with nano-reinforcements is presented. These three-component systems, sometimes called tri-component or multiscale, are intended to give multifunctional properties to the polymer matrix, for instance, outstanding mechanical behavior with increased electrical/thermal conductivity or antimicrobial surface activity.

This research area is still in its infancy and many authors emphasize that the competition between polymer nanocomposites and classical composites, i.e. micro-composites, remains unlikely due to the great development of the latter. For these three-component composites, carbon nanotubes (CNT), especially multi-walled carbon nanotubes (MWCNT), and carbon nanofibers are the main nanofillers used, and studies with nanoclays or other nanoparticles are hard to find in the literature.

Two techniques have been generally used to disperse nanofillers in a polymer resin: sonication and three-roll milling. Furthermore, when three-component systems are being produced, the main focus usually lies on resin-dominated properties of the final composite, particularly, interlaminar shear strength (ILSS), more recently known as short-beam strength.

To the best of our knowledge, the first journal publication on three-component systems (thermoset resin + microfiber + nanofiller) is due to Gojny et al. [10]. In this work, the authors investigated the potential of reinforcing epoxy resins by adding CNT and carbon black (0.1–0.3 %wt.). The nanofillers were dispersed in the resin and, after that, a three-mill roll was employed to apply high shear rate to the mixture. A significant increase (up to 20 %) in ILSS was obtained by adding the nanofillers to a resin/glass-fiber composite produced using the resin transfer molding (RTM) technique. On the other hand, no significant increase was detected in tensile properties, leading the authors to infer that the replacement of fiber-reinforced polymers with nanocomposites is unrealistic.

In the work of Wichmann et al. [19], the addition of 0.3 %wt. of amino-functionalized double-walled carbon nanotubes (DWCNT) and 0.5 %wt. of epoxy-functionalized fumed silica nanoparticles to an epoxy system (without fibers) lead to an increase in toughness of 42 % and 55 %, respectively. The authors also added the carbon nanotubes and the fumed silica to a glass-fiber reinforced epoxy composite produced via RTM expecting to improve interlaminar strength and toughness. The mixture of resin and nanofillers was conducted in a three-roll mill calendar using high shear rate. In addition, the authors applied an electric field to the material after adding the carbon nanotubes in an attempt to align them and to produce an anisotropic material. However, no significant effect was noticed on the measured properties for the composite with 50 %vol. fiber. A similar finding was reported for the composite with 37 %vol. fiber except for interlaminar toughness.

Seyhan et al. [17] studied interlaminar fracture toughness (modes I and II) and interlaminar shear strength of E-glass/CNT/vinyl ester-epoxy resin. The three-roll milling technique was employed to disperse the nanofillers in the resin. Fracture toughness (mode II), was found be about 8 % higher in comparison with the reference laminate, whereas ILSS increased 11 %.

An interesting approach to assess the potential for stress/strain and damage monitoring of composites was proposed by Boeger et al. [1]. An epoxy resin was modified with two different types of CNT and with carbon black to achieve electrical conductivity in glass-fiber reinforced composites produced by RTM. Specimens were then subjected to incremental tensile and fatigue, and ILSS was measured. During the mechanical tests, electrical conductivity was simultaneously monitored and this technique showed high potential for damage and load detection in FRP structures involving a nanocomposite matrix.

Carbon nanofibers (CNF) have also been used to improve interlaminar fracture in composites. Sadeghian et al. [16] employed CNF, with and without surfactant, to produce nanocomposites. The system without surfactant showed too many aggregates which did not allow an efficient infusion by vacuum-assisted resin transfer molding (VARTM). The authors also showed that the addition of 1 %wt. of CNF improved, in about 100 %, mode-I delamination resistance, G_{IC}, of polyester/glass fiber composites.

Another three-component system that has been studied comprises carbon fiber/epoxy/CNT composites produced by RTM or VARTM [12]. These authors found that 0.3 %wt. of CNT added to CF/epoxy did not cause any change in tensile properties. Nevertheless, flexural modulus, strength and strain to break were reported to increase 11.6 %, 18.0 % and 11.4 %, respectively, as compared to the control CF/epoxy composite.

Fan et al. [5] investigated the role of dual scale glass fiber porous media effect on the dispersion state of MWNT during flow. They inferred that, for aggregates larger than 100 μm, the glass mat acted to break down the average aggregate size and, at the same time, it introduced a filtering effect. In a later study, Fan et al. [6] studied the ILSS of traditional glass fiber reinforced epoxy composites enhanced by injecting MWCNT-epoxy suspensions into stationary glass fiber mats. As reported by them, it is difficult to work with more than 0.5 %wt. of MWCNT in epoxy suspensions using traditional VARTM. Because of that, it was suggested the use of another technique, an injection and double vacuum-assisted resin transfer molding (IDVARTM) process to enable working with more concentrated suspensions, keeping constant the fiber content. The samples produced via IDV-ARTM with 0.5 %, 1 % and 2 %wt. of MWCNT lead to an increase in ILSS of 9.7 %, 20.5 % and 33.1 %, respectively.

As introduced above, there are major difficulties associated with the impregnation of a dry fibrous reinforcement using a highly viscous suspension of resin/nanofiller following the most common techniques of the liquid molding family (e.g. RTM, VARTM, infusion), especially for high content of nanofillers or highly packed fibrous systems. In such strict circumstances, an alternative approach to enable processing of these composites comprises the production/use of prepregs, which is discussed next.

2 Prepregs with nanofillers

Prepreg is a term used for "pre-impregnated" fibers used as raw-material to manufacture composites. The fibers, usually arranged in a unidirectional way, are already infiltrated with a pre-determined amount of a resin material, mostly a thermoset polymer. Other additives/components may also have been added to the system. Since the resin is already fully formulated, it must be refrigerated to extend its shelf life. The prepregs are commonly cured by heat in an oven or autoclave, producing a final costly composite of high quality, intended for specialty applications.

The development of three-component prepregs containing nanofillers is quite recent. Godara et al. [9] dispersed 0.5 %wt. of MWCNT, thin-MWCNT and amino-functionalized DWCNT in a DGEBA epoxy resin by high-shear calandering and then spread it on unidirectional carbon fibers in a drumwinder to produce the prepregs which were later autoclaved to produce laminates. The samples showed a systematic decrease in ILSS for all CNT fillers which was attributed to poor interfacial interaction with the matrix, but an increase in ILSS was obtained when the MWCNT-filled epoxy resin was modified with a compatibilizer. The addition of CNT increased crack initiation and propagation energy, indicating crack bridging by the CNT.

In a more recent publication, Godara et al. [8] studied a different route for the incorporation of CNT in epoxy/glass fiber composites where the fibers were drawn through an epoxy-compatible phenoxy-based sizing containing 0.5 %wt. MWCNT. After that, the fibers were embedded with an epoxy resin filled with

MWCNT producing prepregs which were subsequently used to obtain unidirectional composites by winding. It was clearly demonstrated that ILSS of the composites was improved by the presence of CNT either within the matrix or in the fiber sizing, where the latter showed the best results.

Chang [2] demonstrated the synergetic effect of MWCNT, carbon fiber, and glass fiber on static, dynamic mechanical and thermal properties of epoxy composites. The addition of MWCNT (0.75 phr) enhanced flexural strength of GFRP/EP and CFRP/EP in 22.1 % and 9.9 %, respectively. The addition of a higher loading of MWCNT (2.0 phr) increased impact strength of GFRP/EP and CFRP/EP in 44.3 % and 35.9 %, respectively. Their results also showed an increase in fatigue life (MWCNT/GFRP/EP reached 3232 more cycles than MWCNT/CFRP/EP) and the potential of using pristine MWCNT to effectively decrease the coefficient of thermal expansion of epoxy composites.

The preparation of this kind of prepregs, sometimes called nano-prepregs, has also been reported by Chen et al. [3]. The authors incorporated amino-functionalized MWCNT into ordinary carbon fiber/epoxy prepregs. CNT were initially dispersed in acetone and sonicated for 30 min to unravel the nanotubes, which was followed by the addition of epoxy resin and mechanical stirring for 10 h. The CNT-dispersed system was then placed in an impregnation bath with a commercial prepreg fabric and again sonicated for 30 min to allow dispersion of the CNT across the prepreg. The impregnated fabric was then taken to an oven to remove excess solvent. The nano-prepreg contained 40 %wt. carbon fibers and up to 0.75 %wt. CNT, being later molded by hot compression (1000 psi) at 150 °C for 30 min and subjected to post-curing at 100 °C for 12 h. Increase in MWCNT content and MWCNT functionalization enhanced mechanical and electrical properties of the carbon fiber/epoxy system.

Kim and Hahn [11] described a processing method for potential large-scale use for the manufacturing of graphite fiber/epoxy matrix composite with single-walled carbon nanotubes (SWCNT). The SWCNT were scattered over the surface of a graphite/epoxy prepreg by air spraying. Scanning electron microscopy (SEM) showed that SWCNT were uniformly distributed in the plane of the laminate but not in the perpendicular direction due to the low fluidity of the resin during curing. At high SWCNT concentration (> 1.0 %wt.), the combination of a thick band of nanotubes with low fluidity of the resin, caused lack of resin inside of the SWCNT bundles. The following properties showed a positive effect, especially for low SWCNT content: in-plane shear strength, interlaminar fracture toughness (mode I), compressive strength and electrical conductivity. Addition of 2 %wt. SWCNT increased out-of-plane electrical conductivity by 144 %.

There are also several patents reporting on three-component prepregs containing carbon nanotubes. For instance, de Marco and de Villoria [4] reported a new nanoreinforcement prepreg manufacturing method comprised of: mixing the nanoreinforcement with the resin aided by a solvent, evaporation of the solvent, addition of the hardener, and production of the nanoreinforcement mixture prepreg with polyester fiber fabric, cellulose fabric, carbon or glass fiber. In a patent for producing housings for electronic components [15], Honeywell International Inc.

proposed the use of a mixture of nano- and micro-reinforcements, such as CNT, carbon nanofibers and carbon hollow microspheres in epoxy/carbon fiber commercial prepregs.

3 Case study

In this section, the authors will present a case study focusing on three-component composites comprised of CNT, glass-fibers and epoxy resin. First, the materials and equipments used are described, followed by the developed methodology for the deposition of CNT, the RTM processing of the three-component composite and, finally, the characterization of the obtained composites.

3.1 Experimental

The following materials were used: Araldite LY 1316 (Huntsman) epoxy resin (EP), HY1208 (Huntsman) curing agent, BYK (A500 and A560) degassing agents, acetone (99.9 % purity), and glass fiber (GF) plain-weave cloth (Fibertex), 300 g/m^2. The MWCNT used were produced by Chegdu Organic Chemicals Co. Ltd. and supplied with the following characteristics: 5-30 nm external diameter, 1-30 µm length and 85 % purity.

The developed methodology for obtaining the three-component composites is thoroughly described below.

3.1.1 Standard CNT Direct Dispersion in the Resin Via Sonication

The process of direct dispersion of CNT in the resin was carried out following two somewhat standard stages [18]:

(i) Dispersion of CNT into acetone, and
(ii) Dispersion of the epoxy resin into this mixture.

In the first step, 0.75 g of CNT (20 %wt. in relation to epoxy) was dispersed in 60 ml of acetone (0.25 %wt. in relation to epoxy), with sonication for 30 min at 165 W (22 % amplitude), and simultaneous magnetic stirring (FISATOM, model 725A). In the second stage, epoxy resin (300 g) was added to the mixture, followed by dispersion for 40 min at 225 W (30 % amplitude) and simultaneous magnetic stirring. After that, the solvent was removed by heating (60 °C) under vacuum for 1.5 h and simultaneous magnetic stirring. A high-energy ultrasound equipment from Sonics/Vibracell (model VCX 750) was used for CNT untangling.

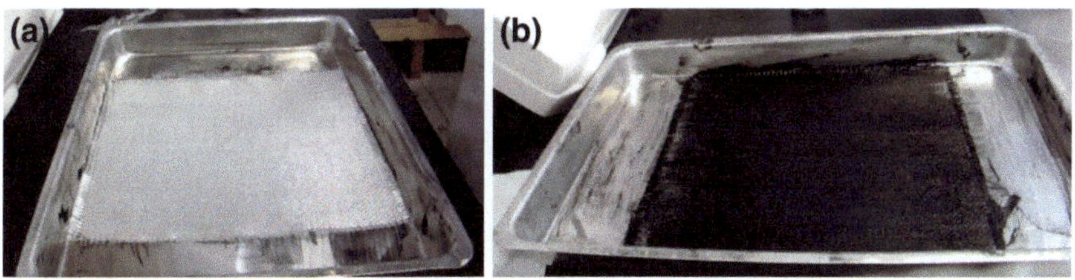

Fig. 1 Visual aspect of the glass-fiber cloth prior to (**a**) and after (**b**) CNT deposition

3.1.2 Development of a Route for CNT Incorporation into Glass-Fiber Cloths

In this study, three attempts for the incorporation of CNT into the glass fiber cloths were made, namely:

(i) Injection of the CNT/acetone suspension using an RTM mold, which was unsuccessful due to filtering of CNT in the fabric close to the injection gate, allowing only the permeation of acetone;
(ii) Direct immersion into a CNT/acetone bath followed by solvent volatilization, which was also unsuccessful due to CNT accumulation on the top layer only, and
(iii) Direct deposition of CNT/acetone: In this route, the CNT were first dispersed in acetone (100 ml per cloth) using high-energy sonication (225 W for 40 min). After that, a small amount of resin was added (15 g per cloth, approx. 3 %wt. in relation to the final composite) and this mixture was sonicated under the same conditions.

Following the third route, the acetone/CNT suspension was deposited over the glass-fiber cloth following a painting-like procedure using a sponge brush, one side at a time, and dried at 80 °C for 120 min in an air-circulation oven to allow thorough solvent volatilization and complete drying of the fibers. Since the suspension had nearly no resin in it, the volatilization condition could be severe, avoiding the deleterious effect of acetone traces in the composite [14]. The process was then repeated on the other side of each cloth prior to RTM molding. This route allowed homogeneous CNT distribution on the cloth as can be verified in Fig. 1.

3.1.3 Molding of Three-Component Composites by RTM

The CNT were deposited (1, 2 and 4 %wt. in relation to composite) over the surface of the glass-fiber cloths using the previously described direct deposition route. In some RTM moldings, CNT were also dispersed in the epoxy resin that was used to infiltrate the reinforcement (0.25 %wt. resin) following the standard previously-mentioned dispersion process. Thus, composites were produced with distinct CNT content, distributed on the cloth and/or within the resin (see Table 1), but keeping constant the glass-fiber volume fraction (~ 30 %vol.).

Table 1 Formulation of the two- and three-component composites produced

Laminates	Dispersed CNT (wt./wt. of resin used for RTM) (%)	Deposited CNT (wt./wt. of fiber used for RTM) (%)	Overall CNT content (wt./wt. of final RTM molded composite) (%)
EP/GF	–	–	–
EP(0.25 %)/GF	0.25	–	0.15
EP/GF(1 % CNT)	–	1.0	1.00
EP(0.25 %)/GF(1 % CNT)	0.25	1.0	1.15
EP/GF(2 % CNT)	–	2.0	2.00
EP(0.25 %)/GF(2 % CNT)	0.25	2.0	2.15
EP/GF(4 % CNT)	–	4.0	4.00
EP(0.25 %)/GF(4 % CNT)	0.25	4.0	4.15

Then, degassing additives were added to the resin, containing or not dispersed CNT, followed by homogenization via mechanical stirring (for 5 min) under vacuum. Immediately prior to RTM molding, the epoxy hardener was added and also homogenized via mechanical stirring (for 5 min). The two- (epoxy/glass fiber) and three-component (epoxy/CNT/glass fiber) composites were molded by radial-infiltration RTM to obtain flat plates ($300 \times 300 \times 2.2$ mm). The whole procedure to obtain the final composites can be visualized in the flowchart depicted in Fig. 2.

The RTM system used consisted of several devices to allow injection and monitoring of the pressure distribution. Observation of the flow behavior with a digital camera was possible through the reinforced-glass top mold. Figure 3 shows the position of the resin flow-front inside the RTM mold after a 2-minute injection time for the different cases. It is easy to verify the difference in resin appearance when the CNT are present in the resin or not (comparing EP(0.25 %)/GF and EP/GF samples, respectively). It can also be noticed the white color of the surface of the non-deposited and yet non-infiltrated glass-fiber cloth (comparing EP/GF and EP(0.25 %)/GF with the other samples).

3.1.4 Characterization of the Composites

Characterization of the produced composites was carried out using mechanical, morphological and electromagnetical analyses. Tensile and flexural testing was carried out in Emic (DL30000) equipment according to ASTM D3039 and D790, respectively, whereas short-beam (ILSS) testing (ASTM D2344 M) was done in Emic (DL 2000) equipment. Izod impact testing (ASTM D256) was performed in CEAST (Impactor II) equipment and the fractured samples were observed using a Scanning Electron Microscope (Jeol JSM 6060).

The Naval Research Laboratories (NRL) arch method was used to perform the electromagnetic characterization by reflectivity measurements using an arch structure. The equipment is located at the Laboratory for Electromagnetic Characterization of the Materials Division of the IAE (Institute of Aeronautics and Space).

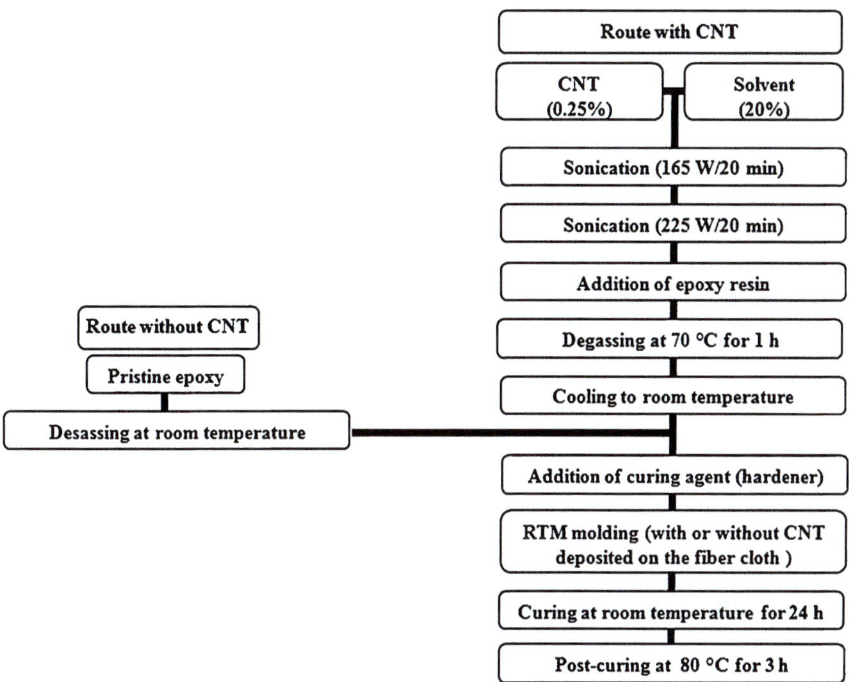

Fig. 2 Flowchart of the whole composite molding procedure

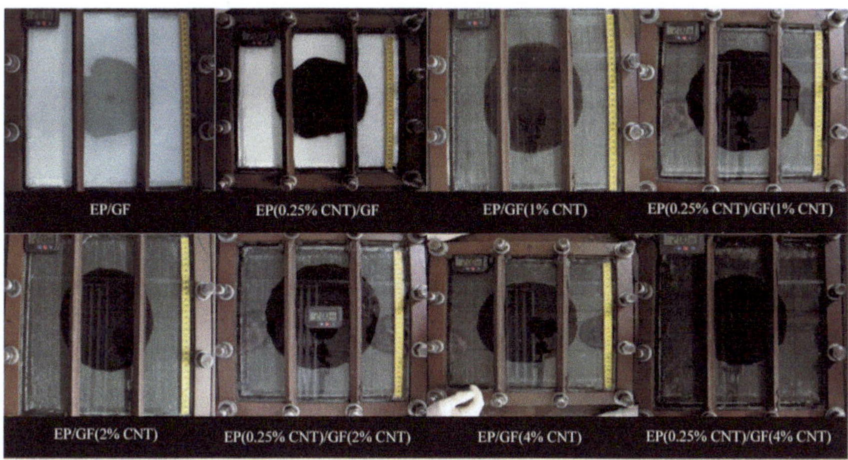

Fig. 3 Aspect of the resin flow-front after 2-min infiltration period

The arch consists of a wooden structure that allows the setting of a pair of antennas in a variety of angles. The horn antenna can be moved along this arch. The sample to be characterized is positioned on a small base at the center of the arch curvature.

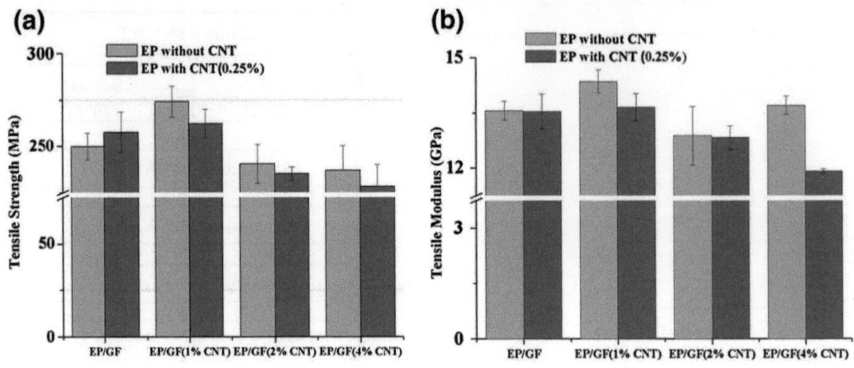

Fig. 4 Tensile strength (**a**) and modulus (**b**) of the composites

The arch structure is designed to keep the horn pointing to the center of the specimen under testing. The transmitter and the receiver can be close to each other as long as an absorbing material is positioned between them in order to reduce interference from the transmitting horn to the receiving horn [7]. The composites were cut in the shape of square plates (dimensions: 200 × 200 mm) and the reflectivity measurements were performed in the 8.2–12.4 GHz frequency range at room temperature. An aluminum metal plate was used as the reference sample. An Anritsu spectrum analyzer (9 kHz to 40 GHz), model MS 2668C, was coupled to the system. The electromagnetic loss under reflection was measured for six of the composites.

3.2 Results and Discussion

3.2.1 Mechanical Analysis

The incorporation of 0.25 %wt. of CNT in the resin did not cause a significant improvement in tensile strength (Fig. 4a) compared to the reference EP/GF composite. In fact, in all cases, the variation in strength was small. Regarding the effect of the deposition of CNT on the cloth, the addition of 1 % CNT appears to lead to an increase in strength (about 9 %), followed by a decrease for higher deposited CNT content, reaching values lower than the reference. This suggests that the deposited CNT may promote adhesion between epoxy and glass-fibers, but only for a controlled CNT amount. Nearly the same trend was found for tensile modulus (Fig. 4b) and only the EP/GF (1 % CNT) sample showed higher stiffness than the reference.

The results of flexural strength and modulus for the composites with varied CNT amounts are shown in Fig. 5. Strength and modulus did not significantly vary, in general, with the introduction of CNT in the EP resin. On the other hand, with the deposition of 1 % CNT on the glass surface, strength increased about 9 %

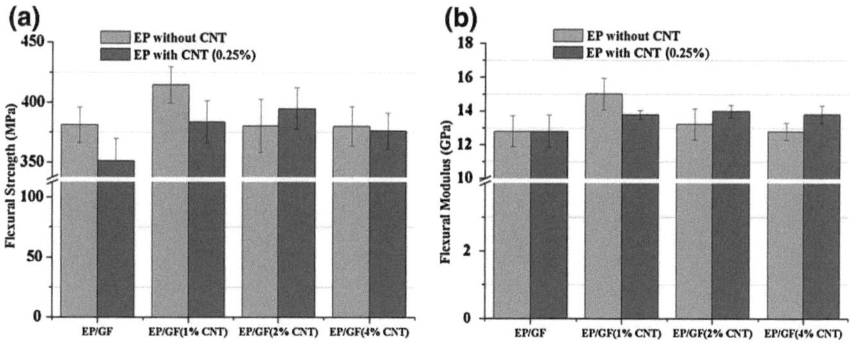

Fig. 5 Flexural strength (**a**) and modulus (**b**) of the composites

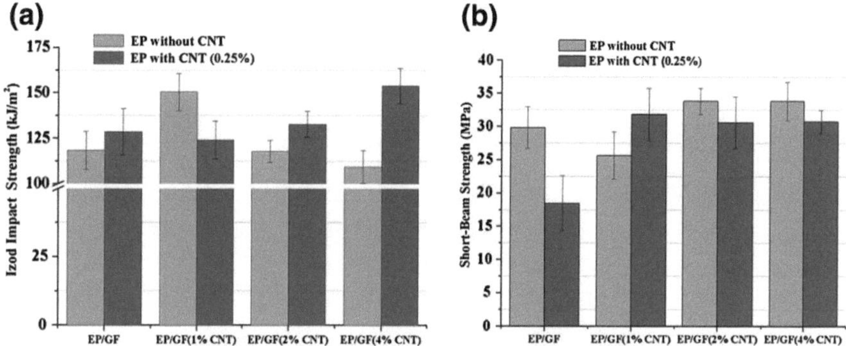

Fig. 6 Impact strength (**a**) and short-beam strength (**b**) of the composites

and modulus about 18 % compared to the reference composite, but only for the resin without CNT. As reported by Zhou et al. [20], a less uniform CNT dispersion may be expected in systems with high CNT loading, which may lead to an increase in void content during processing.

Regarding the impact strength results (Fig. 6a), the effect of the CNT dispersion in the resin is not clear. Nevertheless, the EP/GF(1 % CNT) sample again stands out, which could suggest that the deposition of a large amount of CNT on the cloth may yield under loading, a cohesive failure of the coating, which could behave as an interphase between epoxy and glass-fiber. The highest interlaminar shear strength values (Fig. 6b) were obtained for the EP/GF(2 % CNT) and EP/GF(4 % NTC), but still within the scatter of the results. Fan et al. [6] reported an increase in ILSS with the inclusion of CNT in between glass-fiber cloths, whereas Godara et al. [9] reported a decrease in ILSS for EP/carbon fiber composite with the addition of CNT without surface treatment, which was considered a consequence of the poor CNT/resin interaction.

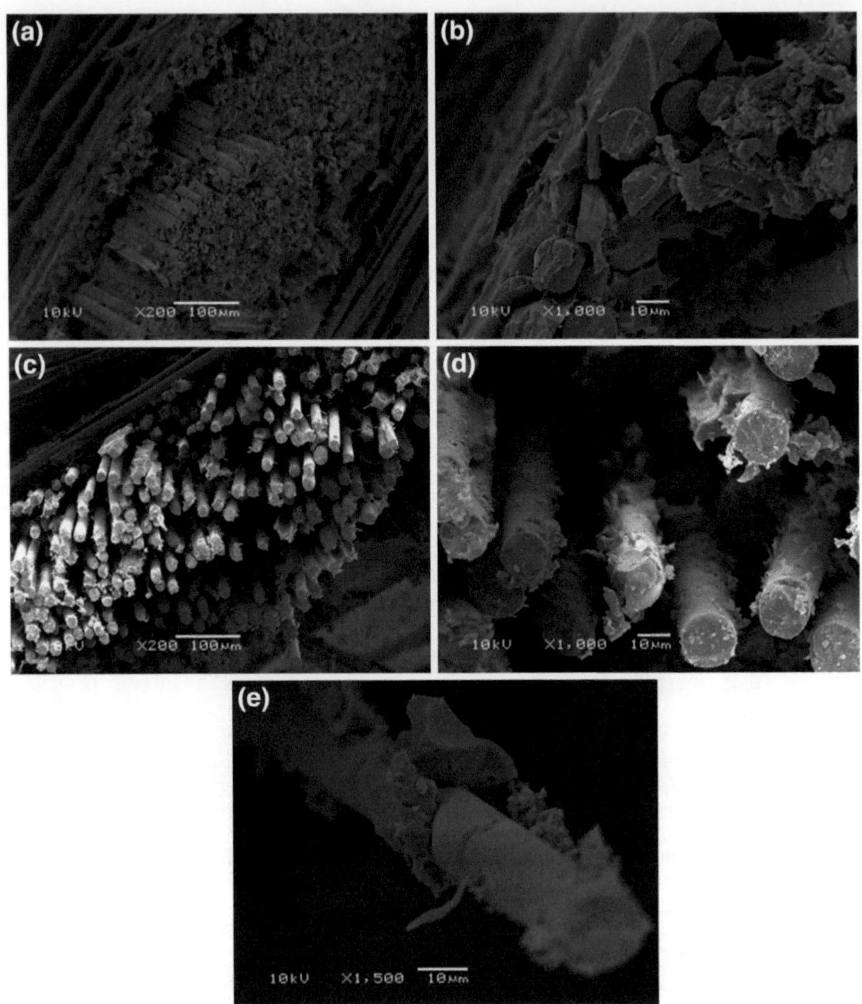

Fig. 7 SEM images of the impact-fractured surface of the EP (0.25 %)/GF sample (**a** and **b**) and of EP/GF (1 % CNT) sample (**c** and **d**), along with a magnified view of the surface of a fiber covered with 1 % CNT (**e**)

3.2.2 Morphological Analysis

The micrographs of the EP(0.25 %)/GF and EP/GF(1 % CNT) samples after impact testing are shown in Fig. 7. In these images, it is possible to observe brittle fracture characteristics as well as somewhat poor adhesion between matrix (with CNT) and fibers for the specimen with 0.25 %wt. of CNT dispersed in the resin (Fig. 7a, b). Indeed, for a better CNT adhesion with the matrix, functionalized CNT should be preferred.

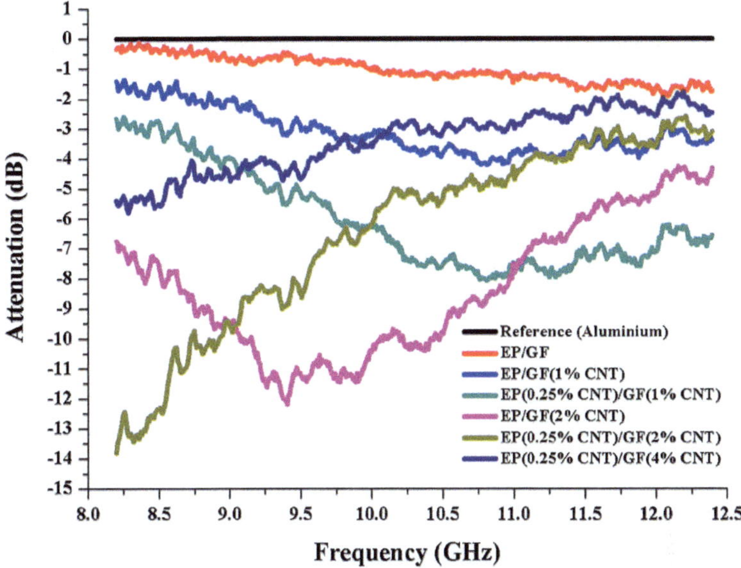

Fig. 8 Attenuation response of the various composites samples in the 8.0–12.5 GHz range of microwave frequency

For the sample containing only CNT deposited on the cloth (Fig. 7c–e), a distinct fracture morphology was obtained, with a large amount of fibers detached from the matrix, showing weaker adhesion than before. From Fig. 7e, which shows a magnified vied of a CNT-covered glass fiber, it is clear that the agglomerated CNT are acting as a reasonably thick coating on the fiber surface, increasing the likelihood of a cohesive failure of this layer and reducing overall strength, as previously shown, especially for the samples with large content of deposited CNT.

3.2.3 Reflectivity Analysis

Reflectivity tests were conducted to determine the response of the composites in the presence of absorbing centers, like the CNT, aiming to use these materials as radiation absorbing materials. The attenuation response of the various samples in the microwave range (8.0–12.5 GHz) of frequency radiation is shown in Fig. 8.

The EP/GF composite showed to be nearly transparent to the electromagnetic waves, basically due to the low electrical conductivity of the glass fibers (mostly composed of silica—SiO_2) [13], being quite similar in behavior to the aluminum plate used as reference. However, there is a tendency towards higher attenuation for higher content of absorbing centers. The maximum value was found for the EP (0.25 % CNT)/GF(2 % CNT) sample, c.a. -14 dB, which represents $\sim$95 % absorption of electromagnetic energy. The EP/GF(4 % CNT) composite did not surpass that probably due to imperfections in the composites originated from irregular coating or wetting.

The EP/GF(2 % CNT) sample also displayed good overall performance in attenuation of incident radiation. It showed maximum attenuation of -12 dB (representing ~ 93 % of electromagnetic absorption) but with a wider attenuation frequency band, especially between 9.0–10.5 GHz.

Thus, attenuation of electromagnetic waves may be developed in EP/GF composites by including CNT as absorbing centers. The CNT induce an electrical current with electromagnetic waves and the flow of current absorbs these waves, as explained by the modified Maxwell's law [13]. This justifies the need to introduce high, yet well-distributed, CNT content, and also the interest in improving the manufacturing process of these high performance materials.

4 Conclusions

In this chapter, a brief review on three-component composites comprised of a (micro)fibrous reinforcement and a thermoset polymer resin filled with nanoparticles is presented. The study of these three-component systems, used to give multifunctional properties to the polymer matrix, is in its infancy and the competition between polymer nanocomposites and classical composites, i.e. microcomposites, is still unlikely due to the large development of the latter. For these composites, carbon nanotubes, especially multi-walled carbon nanotubes, and carbon nanofibers are the most used nanofillers and the focus usually lies on resin-dominated properties of the final composite, such as interlaminar shear strength, and interlaminar fracture toughness.

Many three-component systems comprised of fiber/epoxy/CNT have been produced with RTM or VARTM. However, there are major difficulties associated with the impregnation of a dry fibrous reinforcement using a highly viscous suspension of resin/nanofiller following the most used techniques of the liquid molding family, especially for high content of nanofillers or highly packed fibrous systems. In such strict circumstances, an alternative approach to enable processing comprises the production/use of prepregs. These three-component prepregs containing nanofillers are quite recent in the technical literature (journal papers or patents) and the focus is on a possible synergetic effect between nano-reinforcement and micro-reinforcement, regarding static and dynamic mechanical properties of the composites, electrical conductivity, among others.

From the presented case study, it can be concluded that it was possible to process glass-fiber composites with high content of CNT via resin transfer molding (RTM). A practical, low-cost and effective methodology for the direct deposition of CNT on glass-fiber cloths was developed, achieving up to 4.15 % wt. in overall CNT content in the composite. An acetone/CNT/epoxy suspension was deposited on the glass fiber cloth producing a type of coating on the exposed fibers. The two-step (deposition and molding) methodology followed here is also beneficial considering that an epoxy resin containing only 0.25 %wt. of CNT was used to infiltrate the cloths, enabling RTM to obtain such a high overall CNT content in the final composite.

The tensile, flexural, impact and ILSS strength properties of the composites produced with CNT increased, in general, up to 10 % compared to the reference epoxy/glass-fiber composite. A greater improvement was not noticed because of the particular arrangement of the CNT, which were preferably located on the surface of the glass fibers.

Since most of the acetone/CNT suspension used did not infiltrate the fiber bundles, acting mainly as a coating, a great part of the CNT did not directly interact with the fibers and matrix. However, it is important to add that the mechanical properties were not diminished for a controlled, yet high, amount of CNT content introduced in the composite, which may be of considerable interest when the focus is the improvement of other properties of the final composite, such as electromagnetic shielding.

Indeed, the high CNT content was of uttermost importance for the development of electromagnetic characteristics on the final composite, absorbing much of the radiation in the microwave frequency range. The reflectivity properties of the composites showed an increasing trend for higher CNT content (up to 2.15 %wt.), reaching a maximum value of approximately -14 dB (~ 95 % of electromagnetic absorption). This performance was obtained using a comparatively low cost (glass fiber) and thin (~ 2.2 mm) polymer composite material. Thus, the produced composites showed great potential to be used as microwave-absorption materials, replacing conventional ones employed for this aim. With further improvement in the manufacturing process, these materials could be of interest as high performance composites in a wide range of engineering applications, from telecommunications to aerospace.

Acknowledgments The authors wish to thank Dr. Mirabel Rezende (IAE/CTA) for the reflectivity measurements, Dr. Ademir Zattera (UCS) for the short-beam testing and Giulio Toso for help with the moldings. The authors would also like to thank CNPq, CAPES and FAPERGS for the financial support.

References

1. Boeger, L., Wichmann, M.H.G., Meyer, L.O., Schulte, K.: Load and health monitoring in glass fibre reinforced composites with an electrically conductive nanocomposite epoxy matrix. Compos. Sci. Technol. **68**, 1886–1894 (2008). doi:10.1016/j.compscitech.2008.01.001
2. Chang, M.S.: An investigation on the dynamic behavior and thermal properties of MWCNTs/ FRP laminate composites. J. Reinf. Plast. Comp. **29**, 3593–3599 (2010). doi:10.1177/ 0731684410379510
3. Chen, W.J., Li, Y.L., Chiang, C.L., Kuan, C.F., Kuan, H.C., Lin, T.T., Yip, M.C.: Preparation and characterization of carbon nanotubes/epoxy resin nano-prepreg for nanocomposites. J. Phys. Chem. Solids **71**, 431–435 (2010). doi:10.1016/j.jpcs.2009.12.006
4. De Marco, A.M., De Villoria, R.G.: Manufacturing nanoreinforcement prepreg used for e.g. monitoring damage in structures, electrostatic shielding, anti-lightning systems in wind generators, and surface finishing, involves mixing nanoreinforcement with resin. Patent EP2000494-A1 (2009)

5. Fan, Z.H., Hsiao, K.T., Advani, S.G.: Experimental investigation of dispersion during flow of multi-walled carbon nanotube/polymer suspension in fibrous porous media. Carbon **42**, 871–876 (2004). doi:10.1016/j.carbon.2004.01.067

6. Fan, Z.H., Santare, M.H., Advani, S.G.: Interlaminar shear strength of glass fiber reinforced epoxy composites enhanced with multi-walled carbon nanotubes. Compos. Part A-Appl. S. **39**, 540–554 (2008). doi:10.1016/j.compositesa.2007.11.013

7. Folgueras, L.C., Alves, M.A., Rezende, M.C.: Microwave absorbing paints and sheets based on carbonyl iron and polyaniline: measurement and simulation of their properties. J. Aerosp. Technol. Manag. **2**, 63–70 (2010). doi:10.5028/jatm.2010.02016370

8. Godara, A., Gorbatikh, L., Kalinka, G., Warrier, A., Rochez, O., Mezzo, L., Luizi, F., van Vuure, A.W., Lomov, S.V., Verpoest, I.: Interfacial shear strength of a glass fiber/epoxy bonding in composites modified with carbon nanotubes. Compos. Sci. Technol. **70**, 1346–1352 (2010). doi:10.1016/j.compscitech.2010.04.010

9. Godara, A., Mezzo, L., Luizi, F., Warrier, A., Lomov, S.V., van Vuure, A.W., Gorbatikh, L., Moldenaers, P., Verpoest, I.: Influence of carbon nanotube reinforcement on the processing and the mechanical behaviour of carbon fiber/epoxy composites. Carbon **47**, 2914–2923 (2009). doi:10.1016/j.carbon.2009.06.039

10. Gojny, F.H., Wichmann, M.H.G., Fiedler, B., Bauhofer, W., Schulte, K.: Influence of nano-modification on the mechanical and electrical properties of conventional fibre-reinforced composites. Compos. Part A-Appl. S. **36**, 1525–1535 (2005). doi:10.1016/j.compositesa.2005.02.007

11. Kim, H.S., Hahn, H.T.: Graphite fiber composites interlayered with single-walled carbon nanotubes. J. Compos. Mater. **45**, 1109–1120 (2011). doi:10.1177/0021998311402726

12. Kim, M., Park, Y.-B., Okoli, O.I., Zhang, C.: Processing, characterization, and modeling of carbon nanotube-reinforced multiscale composites. Compos. Sci. Technol. **69**, 335–342 (2009). doi:10.1016/j.compscitech.2008.10.019

13. Lee, O.K.H., Kim, S.S., Lima, Y.S.: Conduction noise absorption by fiber-reinforced epoxy composites with carbon nanotubes. J. Magn. Magn. Mater. **323**, 587–591 (2011). doi:10.1016/j.jmmm.2010.10.018

14. Loos, M.R., Coelho, L.A.F., Pezzin, S.H., Amico, S.C.: The effect of acetone addition on the properties of epoxy. Polimeros **18**, 76–80 (2008). doi:10.1590/S0104-14282008000100015

15. Lui, S.D., Stevenson, J.F., Vacanti, D.C., Vicanti, D.C., Lui, S.C.D.: Multimaterial prepreg sheet for forming electronic chassis and for athletic equipment such as vaulting poles or golf clubs, comprises braid or woven fabric sheet of predetermined length and width. Patent US2009/0095523A1 (2009)

16. Sadeghian, R., Gangireddy, S., Minaie, B., Hsiao, K.T.: Manufacturing carbon nanofibers toughened polyester/glass fiber composites using vacuum assisted resin transfer molding for enhancing the mode-I delamination resistance. Compos. Part A-Appl. S. **37**, 1787–1795 (2006). doi:10.1016/j.compositesa.2005.09.010

17. Seyhan, A.T., Tanoglu, M., Schulte, K.: Mode I and mode II fracture toughness of E-glass non-crimp fabric/carbon nanotube (CNT) modified polymer based composites. Eng. Fract. Mech. **75**, 5151–5162 (2008). doi:10.1016/j.engfracmech.2008.08.003

18. Suave, J., Coelho, L.A.F., Amico, S.C., Pezzin, S.H.: Effect of sonication on thermo-mechanical properties of epoxy nanocomposites with carboxylated-SWNT. Mat. Sci. Eng. A-Struct. **509**, 57–62 (2009). doi:10.1016/j.msea.2009.01.036

19. Wichmann, M.H.G., Sumfleth, J., Gojny, F.H., Quaresimin, M., Fiedler, B., Schulte, K.: Glass-fibre-reinforced composites with enhanced mechanical and electrical properties—benefits and limitations of a nanoparticle modified matrix. Eng. Fract. Mech. **73**, 2346–2359 (2006). doi:10.1016/j.engfracmech.2006.05.015

20. Zhou, Y., Pervin, F., Lewis, L., Jeelani, S.: Experimental study on the thermal and mechanical properties of Multi-walled carbon nanotube-reinforced epoxy. Mat. Sci. Eng. A-Struct. **452**, 657–664 (2007). doi:10.1016/j.msea.2006.11.066